A Course in In-Memory Data Management

Hasso Plattner

A Course in In-Memory Data Management

The Inner Mechanics of In-Memory Databases

Second Edition

 Springer

Hasso Plattner
Enterprise Platform and Integration Concepts
Hasso Plattner Institute
Potsdam
Germany

ISBN 978-3-642-55269-4   ISBN 978-3-642-55270-0 (eBook)
DOI 10.1007/978-3-642-55270-0
Springer Heidelberg New York Dordrecht London

Library of Congress Control Number: 2014940411

Printed on acid-free paper

Springer is part of Springer Science+Business Media (www.springer.com)

Preface

Why We Wrote This Book

Our research group at the Hasso Plattner Institute (HPI) in Potsdam, Germany conducts research in the area of in-memory data management for enterprise applications since 2006. Since then, the ideas and concepts behind dictionary-encoded column-oriented in-memory databases have gained much traction, not only due to the success of SAP HANA as the cutting-edge industry product. As this topic reached a broader audience, we felt the need for proper education in this area. This is of utmost importance as students and software developers have to understand the underlying concepts and technology in order to make most use of it.

At our institute, we have been teaching in-memory data management to students in a Master's course since 2009. When I learned about the current movement towards the direction of Massive Open Online Courses, I immediately decided that we should offer our course about in-memory data management to the public. On September 3, 2012 we launched our online lecture on the new e-learning platform http://www.openHPI.de. Since then, we granted more than 4,500 graded certificates to a total of over 28,000 participating learners. Please feel free to register at openHPI.de to be informed about upcoming lectures and take part in the next online course.

Several thousand people have already used our material in order to study for the homework assignments and final exam of this online course. This book is based on the reading material that we provided to the online community. In addition to that, we incorporated many suggestions for improvement as well as self-test questions and explanations. As a result, we provide you with a textbook teaching you the inner mechanics of a dictionary-encoded column-oriented in-memory database.

Navigating the Chapters

When giving a lecture, content is typically taught in a sequential manner. You have the advantage that you can read the book according to your interests. To help you navigating through the chapters, we provide a learning map showing how the different learning units build up on each other. For example, the learning unit "Differential Buffer" (Chap. 25) is found relatively late in the book. Nevertheless, you might already read it earlier. It requires that you understood the concepts of how "DELETEs", "INSERTs", and "UPDATEs" are conducted without a differential buffer.

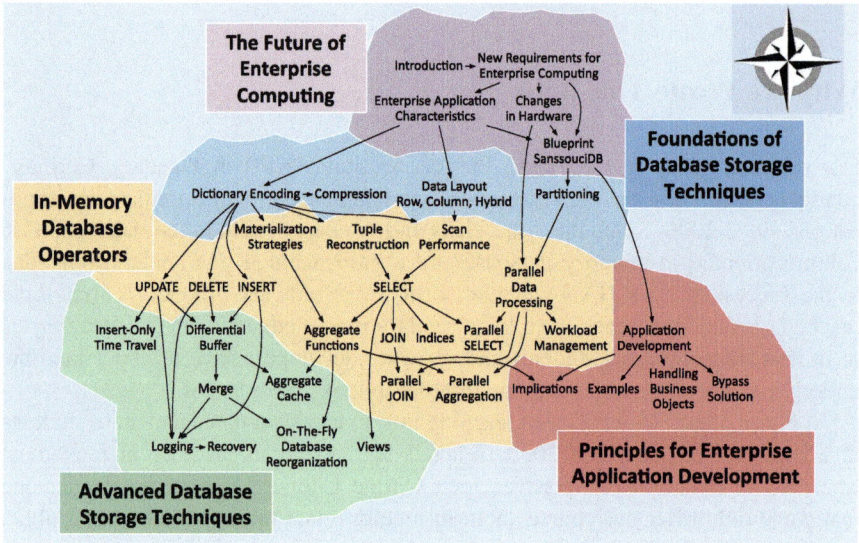

The last section of each chapter contains self-test questions. You also find the questions together with the solutions and explanations at the end of this book.

Writing This Book Is Teamwork

I want to thank the team of my research group "Enterprise Platform and Integration Concepts" at the Hasso Plattner Institute at the University of Potsdam. This book would not exist without this team. Before, during, and after the online lectures, the whole research group took care that no email remained unanswered and feedback from the learning community was incorporated into the learning material.

Among the many contributors who deserve special mention for creating the initial release of this book and its newly revised second edition are *Ralf Teusner*,

Keven Richly, Anja Bog, Martin Boissier, Lars Butzmann, Ralf Diestelkämper, Cindy Fähnrich, Martin Faust, Martin Grund, Franziska Häger, Thomas Kowark, Jens Krüger, Martin Lorenz, Carsten Meyer, Jürgen Müller, Stephan Müller, Matthieu Schapranow, David Schwalb, Christian Schwarz, Christian Tinnefeld, Arian Treffer, Matthias Uflacker, Thomas Uhde, Johannes Wust as well as our team assistant *Andrea Lange*.

Last but not least, I also want to thank our thousands of online learners for their suggestions that resulted in steady improvements in the learning material. We continuously aim to further improve the learning material provided in this book and hope for your valuable feedback. If you identify any flaws, please do not hesitate to contact me at hasso.plattner@hpi.uni-potsdam.de.

We are thankful for any kind of feedback and hope that the learning material will be further improved by the in-memory database community.

Potsdam, Germany   Hasso Plattner

List of Abbreviations

ATP	Available-to-Promise
BI	Business Intelligence
ccNUMA	Cache-Coherent Non-Uniform Memory Access
CPU	Central Processing Unit
DML	Data Manipulation Language
DPL	Data Prefetch Logic
DRAM	Dynamic Random Access Memory
e.g.	for example
EMR	Electronic Medical Records
EPIC	Enterprise Platform and Integration Concepts
ERP	Enterprise Resource Planning
etc.	et cetera
ETL	Extract Transform Load
et al.	and others
FSB	Front Side Bus
HPI	Hasso Plattner Institute
HR	human resources
i.e.	that is
IMC	Integrated Memory Controller
IMDB	In-Memory Database
IO	Index Offsets
IP	Index Positions
MDX	Multidimensional Expression
MIPS	Million Instructions Per Second
NGS	Next-Generation Sequencing
NUMA	Non-Uniform Memory Access
OLAP	Online Analytical Processing
OLTP	Online Transaction Processing
ORM	Object-Relational Mapping
PADD	Parallel Add
PDA	Personal Digital Assistant

QPI	Quick Path Interconnect
RAM	Random Access Memory
RISC	Reduced Instruction Set Computing
SADD	Scalar Add
SIMD	Single Instruction Multiple Data
SRAM	Static Random Access Memory
SSE	Streaming SIMD Extensions
TLB	Translation Lookaside Buffer
UMA	Uniform Memory Access

Contents

Part IV   Advanced Database Storage Techniques

List of Figures

List of Tables

Chapter 1
Introduction

This book *In-Memory Data Management* focuses on the technical details of in-memory columnar databases. Over the last years, in-memory databases and especially column-oriented in-memory databases grew in popularity and became widely researched [BMK09, KNF$^+$12, Pla09]. With modern hardware technologies and increasing main memory capacities, this approach to data management promises groundbreaking new applications.

1.1 Goals of the Lecture

Everybody who is interested in the future of databases and enterprise data management should benefit from this course, regardless whether still studying, already working, or perhaps even developing software in the affected fields. The primary goal of this course is to achieve a deep understanding of column-oriented, dictionary-encoded in-memory databases and the implications of those for enterprise applications. This learning material does not include introductions into Structured Query Language (SQL) or similar basics; these topics are expected to be prior knowledge. However, even if you do not yet have solid SQL knowledge, we encourage you to follow the course since most examples with relation to SQL will be understandable from the context.

With new applications and upcoming hardware improvements, fundamental changes will take place in enterprise applications. The participants ought to understand the technical foundation of next generation database technologies and get a feeling for the difference between in-memory databases and traditional databases on disk. In particular, you will learn why and how these new technologies enable performance improvements by factors of up to 100,000.

H. Plattner, *A Course in In-Memory Data Management*,
DOI 10.1007/978-3-642-55270-0_1, © Springer-Verlag Berlin Heidelberg 2014

1.2   The Idea

The foundation for the learning material is an idea that professor Hasso Plattner and his "Enterprise Platform and Integration Concepts" (EPIC) research group came up with in a discussion in 2006. At this time, lectures about Enterprise Resource Planning (ERP) systems were rather dry with no intersections to modern technologies as used by Google, Twitter, Facebook, and several others.

The team decided to start a new radical approach for ERP systems. To start from scratch, the particular enabling technologies and possibilities of upcoming computer systems had to be identified. With this foundation, they designed a completely new system based on two major trends in hardware technologies:

- Massively parallel systems with an increasing number of Central Processing Units (CPUs) and CPU-cores
- Increasing main memory volumes

To leverage the parallelism of modern hardware, substantial changes had to be made. Current systems were already parallel in respective to their ability to handle thousands of concurrent users. However, the underlying applications were not utilizing parallelism.

Making use of hardware parallelism is difficult. Hennessy et al. [PH12] discuss what changes have to be made to make an application run in parallel, and explain why it is often very hard to change sequential applications to use multiple cores efficiently.

For the first prototypes, the team decided to look more closely into accounting systems. In 2006 computers were not yet capable of keeping big companies' data completely in memory, and so the decision was made to concentrate on rather small companies in the first step. It was clear that the progress in hardware development would continue and that the advances will automatically enable the systems to keep bigger volumes of data in memory.

Another important design decision was the complete removal of materialized aggregates. In 2006 ERP systems were highly depending on pre-computed aggregates. With the computing power of upcoming systems, the new design was not only capable of increasing the granularity of aggregates, but of completely removing them.

As the new system keeps every bit of the processed information in memory, disks are only used for archiving, backup, and recovery. The primary persistence is the Dynamic Random Access Memory (DRAM), which is accomplished by increased capacities and data compression.

To evaluate the new approach, several Bachelor and Master projects implemented new applications using in-memory database technology over the course of the next years. Ongoing research focuses on the most promising findings of these projects as well as completely new approaches to enterprise computing with an enhanced user experience in mind.

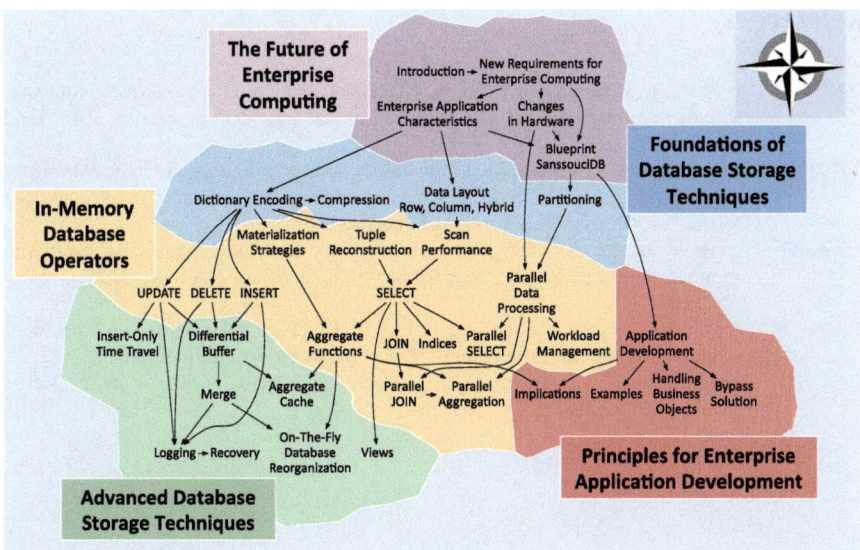

Fig. 1.1 Learning map

1.3   Learning Map

The learning map (Fig. 1.1) gives an overview over the parts of the learning material and the respective chapters in these parts. In this graph, you can easily see what the prerequisites for a chapter are and which contents will follow.

1.4   Self Test Questions

1. **Rely on Disks**
   Does an in-memory database still rely on disks?

   (a) Yes, because disk is faster than main memory when doing complex calculations
   (b) No, data is kept in main memory only
   (c) Yes, because some operations can only be performed on disk
   (d) Yes, for archiving, backup, and recovery

References

[KNF⁺12]   A. Kemper, T. Neumann, F. Funke, V. Leis, H. Mühe, Hyper: adapting columnar main-memory data management for transactional and query processing. IEEE Data Eng. Bull. **35**(1), 46–51 (2012)

[PH12]   D.A. Patterson, J.L. Hennessy, *Computer Organization and Design: The Hardware/ Software Interface*. The Morgan Kaufmann Series in Computer Architecture and Design, rev. 4th edn. (Academic, San Francisco, 2012)

[BMK09]   A. Peter, Boncz, S. Manegold, M.L. Kersten, Database architecture evolution: mammals flourished long before dinosaurs became extinct. Proc. VLDB **2**(2), 1648–1653 (2009)

[Pla09]   H. Plattner, A common database approach for OLTP and OLAP using an in-memory column database, in *Proceedings of the SIGMOD International Conference on Management of Data*, ed. by U. Çetintemel, S.B. Zdonik, D. Kossmann, N. Tatbul (ACM, 2009), pp. 1–2

Part I
The Future of Enterprise Computing

Chapter 2
New Requirements for Enterprise Computing

When thinking about developing a completely new database management system for enterprise computing we first need to clarify whether there are significant opportunities for improvement in the area. The fact that modern companies have changed dramatically towards a more data-driven model gives the strong indication that this might be the case. For a typical enterprise computing use-case like assembly line management it now became viable to use sensors giving instant feedback on various parameters and producing huge amounts of data. Furthermore companies process previously available data at a much larger scale, e.g. competitor behavior, price trends, etc. to support management decisions. All this data volume will continue to increase in the future.

There are two major requirements for a modern database management system:

- Data from various sources need to be combined in a single database management system, and
- This data needs to be analyzed in real-time to support interactive decision-making.

The following sections outline typical use cases for modern enterprises and derive associated requirements for an entirely new enterprise data management system.

2.1 Processing of Event Data

Event data influences enterprises more and more today. It is characterized by the following aspects:

- Each event dataset itself is small (some bytes or kilobytes) compared to the size of traditional enterprise data, such as all data contained in a single sales order, and

H. Plattner, *A Course in In-Memory Data Management*,
DOI 10.1007/978-3-642-55270-0_2, © Springer-Verlag Berlin Heidelberg 2014

- The number of generated events for a specific entity is high compared to the amount of entities, e.g. hundreds or thousand events are generated for a single product.

In the next sections we will give some examples use-cases related to the processing of event data in modern enterprises.

2.1.1   Sensor Data

Sensors are used to supervise the function of more and more systems today. One example is the tracking and tracing of sensitive goods, such as pharmaceuticals, clothes, or spare parts. Hereby packages are equipped with Radio-Frequency Identification (RFID) tags or two-dimensional bar codes, the so-called data matrix. Each product is virtually represented by an Electronic Product Code (EPC), which describes the manufacturer of a product, the product category, and a unique serial number. As a result, each product can be uniquely identified by its EPC code. In contrast, traditional one-dimensional bar codes can only be used for identification of classes of products due to their limited domain set. Once a product passes through a reader gate, a reading event is captured. The reading event consists of the current reading location, timestamp, the current business step, e.g. receiving, unpacking, repacking or shipping, and further related details. All events are stored in decentralized event repositories.

Real-Time Tracking of Pharmaceuticals

For example, approximately 15 billion pharmaceuticals are produced in Europe that are only available on prescription. Tracking any of them results in approx. 8,000 read event notifications per second. These events build the basis for anti-counterfeiting techniques. For example, the route of a specific pharmaceutical can be reconstructed by analyzing all relevant reading events. The in-memory technology enables tracing of 10 billion events in less than 100 ms.

Formula One Racing Cars

Formula one racing cars are also generating excessive sensor data. These sports cars are equipped with up to 600 individual sensors, each recording tens to hundreds of events per second. Capturing sensor data for a 2 h race produces giga- or even terabytes of sensor data depending on their granularity. The challenge is to capture, process, and analyze the acquired data during the race to optimize the car parameters instantly, e.g. to detect part faults, optimize fuel consumption or top speed.

2.1.2 Analysis of Game Events

Personalized content in online games is a success factor for the gaming industry. Browser games can generate a steady stream of tens of thousands events per second, such as player movements, transfer of virtual goods, or general game statistics. Traditional databases do not support processing of these huge amounts of data in an interactive way, e.g. join and full table scans require complex index structures or data warehouse systems optimized to return some selected aspects in a very fast way. However, individual and flexible queries from developers or marketing experts cannot be answered interactively.

Gamers tend to spend money when virtual goods or promotions are provided in a critical game state, e.g. a lost adventure or a long-running level that needs to be passed. In-game trade promotion management needs to analyze the user data, the current in-game events, and external details, e.g. current discount prices.

In-memory database technology is used to conduct in-game trade promotions and, at the same time, conduct A/B testing. To this end, the gamers are divided into two segments. The promotion is applied to one group. Since the feedback of the users is analyzed in real-time, the decision to roll-out a huge promotion can be taken within seconds after the small test group accepted the promotion.

Furthermore, in-memory technology improves discovery of target groups and testing of beta features, real-time prediction, and evaluation of advert placement.

2.2 Combination of Structured and Unstructured Data

First we want to understand structured data as any kind of data that is stored in a format which is automatically processed by computers. Examples for structured data are ERP data stored in relational database tables, tree structures, arrays, etc. Following that we want to understand partially or mostly unstructured data that cannot easily be processed automatically, e.g. all data that is available as raw documents, such as videos or photos. In addition, any kind of unformatted text, such as freely entered text in a text field, document, spreadsheet, or database, is considered as unstructured data unless a data model for its interpretation is available, e.g. a possible semantic ontology.

For years enterprise data management focused on structured data only. Structured data is stored in a relational database format using tables with specific attributes. However, many documents, papers, reports, web sites, etc. are only available in an unstructured format, e.g. text documents. Information within these documents is typically identified via the document's meta data. A detailed search within the content of these documents or the extraction of specific facts is however not possible by using the meta data. As a result, there is a need to harvest information buried within unstructured enterprise data. Searching any kind of data—structured or unstructured—needs to be equally flexible and fast.

2.2.1 Patient Data

In the course of the patient treatment process, e.g. in hospitals, structured and unstructured data is generated. Examples of unstructured data are diagnosis reports, histologies, and tumor documentations. Examples of structured data are results of the erythrogram, blood pressure, temperature measurements, or the patient's gender. The in-memory technology enables the combination of both classes of patient data with additional external sources, such as clinical trials, pharmacological combinations or side-effects. As a result, physicians can prove their hypotheses by interactively combing data and reduce necessary manual and time-consuming searches. Physicians are able to access all relevant patient data and to take their decision on latest available patient details.

Due to their high fluctuation of unexpected events, such as emergencies or delayed surgeries, the daily time schedule of physicians is very time-optimized. In addition to certain technical requirements of their tools, they have also very strict response time requirements. For example, the HANA Oncolyzer, an application for physicians and researchers was designed for mobile devices. The mobile application supports the use-as-you-go factor, i.e., the required patient data is available at any location on the hospital campus and the physician is no longer forced to go to a certain desktop computer for checking a certain aspect. In addition, if the required detail is not available in real-time for the physician, she/he will no longer use the application. Thus, all analyses performed by the in-memory database are running on a server landscape in the IT department while the mobile application is the remote user interface for it.

Having the flexibility to request arbitrary analyses and getting the results within milliseconds back to the mobile application makes in-memory technology a perfect technology for the requirements of physicians. Furthermore, the mobility aspect bridges the gap between the IT department where data is stored and the physician that visits multiple work places throughout the hospital every day.

2.2.2 Airplane Maintenance Reports

Airport maintenance logs are documented during exchange of any spare parts. These reports contain structured data, such as date and time of the replacement or order number of the spare part, and unstructured data, e.g. kind of damage, location, and observations in the spacial context of the part. By combining structured and unstructured data, in-memory technology supports the detection of correlations, e.g. how often a specific part was replaced in a specific aircraft or location. As a result, maintenance managers are able to discover risks for damages before a certain risk for human-beings occurs.

2.3   Social Networks and the Web

Social networks are very popular today. Meanwhile, the time when they were only used to update friends about current activities are long gone. Nowadays, they are also used by enterprises for global branding, marketing and recruiting.

Additionally, they generate a huge amount of data, e.g. Twitter deals with one billion new tweets in five days. This data is analyzed, e.g. to detect messages about a new product, competitor activities, or to prevent service abuses. Combining social media data with external details, e.g. sales campaigns or seasonal weather details, market trends for certain products or product classes can be derived. These insights are valuable, e.g. for marketing campaigns or even to control the manufacturing rate.

Another example for extracting business relevant information from the Web is monitoring search terms. The search engine Google analyzes regional and global search trends. For example, searches for "influenza" and flu related terms can be interpreted as a indicator for a spread out of the influenza disease. By combining location data and search terms, Google is able to draw a map of regions that might be affected from an influenza epidemic.

2.4   Operating Cloud Environments

Operating software systems in the cloud require a good data integration strategy. Assume, you process all your company's human resources (HR) tasks in an on-demand HR system provided by provider A. Consider a change of the provider to cloud provider B. Of course, a standardized data format for HR records can be used to export data from A and import it into B. However, what happens if there is no compatible standard for your application? Then, the data exported from A needs to be migrated, respectively remodeled, before it can be imported into B. Data transformation is a complex and time-consuming task which often has to be done manually due to the required knowledge about source and target formats and many exceptions which have to be solved separately.

In-memory technology provides a transparent view concept. Views describe how input values are transformed to the desired output format. The required transformations are performed automatically when the view is called. For example, consider the attributes `first name` and `last name` that need to be transformed into a single attribute `contact name`. A possible view `contact name` performs the concatenation of both attributes by performing `concat(first name, last name)`.

Thus, in-memory technology does not change the input data, while offering the required data formats by transparent processing of the view functions. This enables a transparent data integration compared to the traditional Extract Transform and Load (ETL) process used for Business Intelligence (BI) systems.

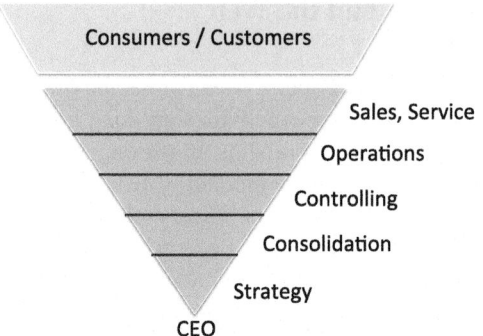

Fig. 2.1 Inversion of corporate structures

2.5   Mobile Applications

The wide spread of mobile applications fundamentally changed the way enterprises process information. First (BI) systems were designed to provide detailed business insights for CEOs and controllers only. Nowadays, every employee is gaining insights by the use of BI systems. However, for decades information retrieval was bound to stationary desktop computers. With the wide-spread of mobile devices, e.g. PDAs, smartphones, etc., even field workers are able to analyze sales reports or retrieve the latest sales funnel for a certain product or region.

Figure 2.1 depicts the new design of BI systems, which is no longer top-down but bottom-up. Modern BI systems provide all required information to sales representatives directly talking to customers. Thus, customers and sales representatives build the top of the inverted pyramid.

In-memory databases build the foundation for this new corporate structure. On mobile devices, people are eager to get a response within a few seconds [Oul05, OTRK05, RO05]. With the ability to perform complex and freely formulated queries with a sub-second response, in-memory databases can revolutionize the way employees communicate with customers. An example of the radical improvements through in-memory databases is the dunning run. A traditional dunning process took 20 min on an average SAP system, but by rewriting the dunning run on in-memory technology it now takes less than one second.

2.6   Production and Distribution Planning

Two further prominent use cases for in-memory databases are complex and long-running processes such as production planning and availability checking.

2.6.1 Production Planning

Production planning identifies the current demand for certain products and consequently adjusts the production rate. It analyzes several indicators, such as the users' historic buying behavior, upcoming promotions, stock levels at manufacturers and whole-sellers. Production planning algorithms are complex due to required calculations, which are comparable to those found in BI systems. With an in-memory database, these calculations are now performed directly on latest transactional data. Thus, algorithms are more accurate with respect to current stock levels or production issues, allowing faster reactions to unexpected incidents.

2.6.2 Available-to-Promise Check

The Available-to-Promise (ATP) check validates the availability of certain goods. It analyzes whether the amount of sold and manufactured goods are in balance. With raising numbers of products and sold goods, the complexity of the check increases. In certain situations it can be advantageous to withdraw already agreed goods from certain customers and reschedule them to customers with a higher priority. ATP checks can also take additional data into account, e.g. fees for delayed or canceled deliveries or costs for express delivery if the manufacturer is not able to send out all goods in time.

Due to the long processing time, ATP checks are executed on top of pre-aggregated totals, e.g. stock level aggregates per day. Using in-memory databases enables ATP checks to be performed on the latest data without using pre-aggregated totals. Thus, manufacturing and rescheduling decisions can be taken on real-time data. Furthermore, removing aggregates simplifies the overall system architecture significantly whilst adding flexibility.

2.7 Mathematical and Scientific Applications

Mathematics is the most powerful, omnipresent tool we have. It can be used to put a man on the moon or to calculate the numbers of atoms in a molecule. It can be found in the geometric transformations governing the canons of J.S. Bach or in the catalogue of symmetries in Islamic art. Nowadays the field of applied mathematics and scientific computing is heavily dependent on computer aided data processing due to the ever-growing amounts of data. Researchers and scientists ideally want to process such big amounts of data as fast as possible and see their results in seconds—not hours or days—in order to decide how to proceed. As mentioned throughout the book, these requirements are in the context of business data processing addressed by using an in-memory database system: in this section we want to demonstrate that the advancements in the business data processing sector are also beneficial for mathematical applications. Although current mathematical and scientific applications are not directly in the focus of enterprise computing, they may pave the way for tomorrow's new business segments.

2.7.1 Proteomics Research

Proteomics research focuses on the study of the proteome, a term referring to the totality of proteins expressed by a genome, cell, tissue or organism at a given time. The proteome is highly dynamic, changing constantly as a response to the needs and state of the organism. Factors such as cancer or other diseases, can also change the composition of the proteome. Thus, analysis of these changes can be used i.e. for cancer diagnosis.

The prominent way to analyse the (human) proteome is by means of mass spectrometry, a technique that measures the masses and concentration of peptides in a biological sample and exports the data in form of a spectrum. Mass spectrometer emit their measurements as raw data then processed in an analysis pipeline consisting of numerous preprocessing (smoothing, filtering) and analysis steps. To identify signals in the proteome for cancer, multiple samples from different cohorts derived in clinical studies have to be compared, which adds selection and join operations to access cohort data from clinical information systems to the pipeline.

Depending on the approach, the analysis pipeline for proteomics mass spectrum data represents a data-flow program, where each step applies mathematical transformations on the data flowing through the pipeline. Each operator step can be parametrized and replaced with a different implementation or method to solve the transformation, e.g. different algorithms for minimization problems can be applied, or different model types for classifications can be embedded into the pipeline.

Proteomics research benefits from the integration of relational database operators and arbitrary mathematical operations in SanssouciDB, allowing the proteomics researcher to compose complex analysis pipelines in one environment using clinical patient data as well as raw spectrum data inside one database. Since every parameter and algorithm influences the accuracy of the resulting statistical model, iterative tuning of such pipelines in real-time is a requirement in this field: here the fast traversal of data and algorithm execution is crucial.

2.7.2 Graph Data Processing

As motivated in the previous subsection, graph processing presents a major challenge for modern database management systems: there are two very different access types that are typically supported by different storage engines. Graphs are a generic representation suitable for almost any information. Even relational databases and the data stored in tables are basically graphs of related nodes with attributes.

The most common operations on graphs are twofold: On the one hand is graph exploration with the goal to traverse and explore singular paths and on the other hand is graph analytics trying to explore and analyze the whole graph or multiple instances of a similar graph.

In native graph databases explorative traversals are directly executed on a graph structure while in in-memory databases multi-way joins are required to follow a single path. However, the same is true for analytical queries. The advantage of a modern in-memory database is that multiple explorative graph queries can be executed at once using joins and aggregations. For analytical queries, the queries can leverage the capabilities of the parallel join execution engine inside SanssouciDB.

Compared to disk-based databases, in-memory column stores have the advantage of fast scanning and data traversal of the necessary attributes which make it possible to execute complex queries. Depending on the application use case it is not always advisable to use relational databases for graph processing. Especially breadth-first and depth-first graph search algorithms can become very expensive due to repetitive joins. On the other hand there are algorithms to resemble graph structures that can be executed very efficiently, like aggregation on hierarchical tree structures used for bill of materials explosion.

2.7.3 SanssouciDB Data Scientist

The SanssouciDB Data Scientist tool supports users who want to implement and interact with complex analysis pipelines that go beyond standard analytical operations consisting of complex mathematical and machine learning operators orchestrated to complex pipelines.

The tool provides a coherent development and modeling environment integrating the graphical data mining modeling paradigm and the source code development paradigm in one framework. The user can graphically design data flow programs to define her analysis pipeline consisting of relational, domain specific native operators, as well as custom operators which e.g. can be directly implemented by the user in the R programming language for statistical computing.

Modeled pipelines are compiled on the fly and executed on SanssouciDB's calculation engine leveraging data locality and transparent parallel execution. Users can interactively manipulate their pipeline, by tuning operator parameters, exchanging complete operators or manipulate code of custom operators within the tool, while the pipeline is compiled to SanssouciDB on the fly, and thereby showing the resulting effects immediately.

2.8   Self Test Questions

1. Data explosion

Consider the formula 1 race car tracking example, with each race car having 512 sensors, each sensor records 32 events per second whereby each event is 64 byte in size.

How much data is produced by a F1 team, if a team has two cars in the race and the race takes two hours?

Please use the following unit conversions: 1,000 byte = 1 KB, 1,000 KB = 1 MB, 1,000 MB = 1 GB.

(a) 14 GB
(b) 15.1 GB
(c) 32 GB
(d) 7.7 GB

References

[OTRK05]   A. Oulasvirta, S. Tamminen, V. Roto, J. Kuorelahti, Interaction in 4-second bursts: the fragmented nature of attentional resources in mobile hci, in *Proceedings of the SIGCHI Conference on Human Factors in Computing Systems, CHI '05* (ACM, New York, 2005), pp. 919–928

[Oul05]   A. Oulasvirta, The fragmentation of attention in mobile interaction, and what to do with it. Interactions **12**(6), 16–18 (2005)

[RO05]   V. Roto, A. Oulasvirta, Need for non-visual feedback with long response times in mobile hci, in *Special Interest Tracks and Posters of the 14th International Conference on World Wide Web, WWW '05* (ACM, New York, 2005), pp. 775–781

Chapter 3
Enterprise Application Characteristics

3.1 Enterprise Data Sources

An enterprise data management system should be able to handle data coming from several different data sources. In the ecosystem of modern enterprises, many applications work on and produce structured data. Enterprise Resource Planning (ERP) systems, for example, typically create transactional data to capture the operations of a business. More and more event and stream data is created by modern manufacturing machines and sensors. At the same time, large amounts of unstructured data is captured from the web, social networks, log files, support systems, and others.

Business users need to query these different data sources as fast as possible to derive business value from the data or coordinate the operations of the enterprise. Real-time analytics applications may work on any of the data sources and combine them. They analyze the structured data of ERP systems for real-time transactional reporting, classical analytics, planning, and simulation. The data from other data sources can be taken into account as well, for example, a text analytics application can combine a customer sentiment analysis on social network data with sales numbers or a production planning application takes sensor data of the RFID sensors into account.

3.2 OLTP vs. OLAP

An enterprise data management system should be able to handle transactional and analytical query types, which differ in several dimensions. Typical queries for Online Transaction Processing (OLTP) can be the creation of sales orders, invoices, accounting data, the display of a sales order for a single customer, or the display of customer master data. Online Analytical Processing (OLAP) is used to summarize data, often for management reports. Typical analytical reports are for example the sales figures aggregated and grouped by regions, different timeframes and products or the calculation of Key Performance Indicators (KPIs).

Because it has always been considered that these query types are significantly different, the data management systems were split into two separate systems to tune the data storage and schemas accordingly. In the literature, it is claimed that OLTP workloads are write-intensive, whereas OLAP-workloads are read-mostly and that the two workloads rely on "Opposing Laws of Database Physics" [Fre95].

Yet, data and query analysis of current enterprise systems showed that this statement is not true [KGZP10, KKG+11]. The main difference between systems that handle these query types is that OLTP systems handle more queries with a single select or queries that are highly selective returning only a few tuples, whereas OLAP systems calculate aggregations for only a few columns of a table, but for a large number of tuples.

For the synchronization of the analytical system with the transactional system(s), a cost-intensive ETL (Extract-Transform-Load) process is required. The ETL process introduces a delay and is relatively complex, because all relevant changes have to be extracted from the outside source or sources if there are several, data is transformed to fit analytical needs, and it is loaded into the target database.

3.3   Drawbacks of the Separation of OLAP from OLTP

While the separation of the database into two systems allows for workload specific optimizations in both systems, it also has a number of drawbacks:

- The OLAP system does not have the latest data, because the ETL process introduces a delay. The delay can range from minutes to hours, or even days. Consequently, many decisions have to rely on stale data instead of using the latest information.
- To achieve acceptable performance, OLAP systems work with predefined, materialized aggregates which reduce the query flexibility of the user.
- Data redundancy is high. Similar information is stored in both systems, just differently optimized.
- The schemas of the OLTP and OLAP systems are different, which introduces complexity for applications using both of them and for the ETL process synchronizing data between the systems.

3.4   The OLTP vs. OLAP Access Pattern Myth

Transactional processing is often considered to have an equal share of read and write operations, while analytical processing is dominated by large reads and range queries. However, a workload analysis of multiple real customer systems reveals that OLTP and OLAP systems are not as different as expected in classic enterprise

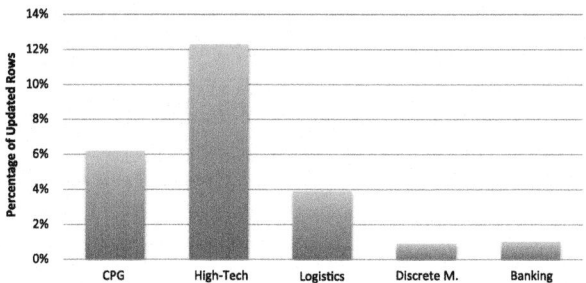

Fig. 3.1 Updates of a financial application in different industries

systems. As shown in [KKG$^+$11], even OLTP systems process over 80 % of read queries, and only a fraction of the workload contains write queries. Less than 10 % of the actual workload are queries that modify existing data, e.g. updates and deletes. OLAP systems process an even larger amount of read queries, which make up about 95 % of the workload.

The updates in the transactional workload are of particular interest. An analysis of the updates in different industries is shown in Fig. 3.1. It confirms that the number of updates in OLTP systems is quite low [KKG$^+$11], and varies between industries. In the analyzed high-tech companies, the update rate peaks at about 12 %, meaning that about 88 % of all tuples saved in the transactional database are never updated. In other sectors, research showed even lower update rates, e.g., less than 1 % in banking and discrete manufacturing [KKG$^+$11].

These results lead to the assumption that updates and deletes can be implemented by inserting tuples with timestamps to record the period of their validity.

The additional benefit of this so called insert-only approach is that the complete transactional data history and a tuple's life cycle are saved in the database automatically, avoiding the need for log-keeping in the application for auditing reasons. More details about the insert-only approach will be provided in Chap. 26.

We can conclude that the characteristics of the two workload categories are not that different after all, which leads to the vision of reuniting the two systems and to combine OLTP and OLAP data in one system.

3.5 Combining OLTP and OLAP Data

The main benefit of the combination is that both, transactional and analytical queries can be executed on the same database using the same set of data as a "single source of truth". Thereby, the costly ETL process becomes obsolete and all queries are performed against the latest version of the data.

In this book we show that with the use of modern hardware we can eliminate the need for pre-computed aggregates and materialized views. Data aggregation can be executed on-demand and analytical views can be provided without delay. With the expected response time of analytical queries below 1 s, it is possible to perform analytical query processing on the transactional data directly, anytime and from any device. By dropping the pre-computation of aggregates and materialization of views, applications and data structures can be simplified. The management of aggregates and views (building, maintaining, and storing them) is not necessary any longer.

The resulting mixed workload combines the characteristics of OLAP and OLTP workloads. A part of the queries in the workload performs typical transactional request like the selection of a few, complete rows. Others aggregate large amounts of the transactional data to generate real-time analytical reports on the latest data. Especially applications that inherently use access patterns from both workload groups and need access to the up-to-date data benefit greatly from fast access to large amounts of transactional data, e.g. dunning or planning applications.

More application examples are given in Chap. 35.

3.6   Enterprise Data is Sparse Data

By analyzing enterprise data in standard software, special data characteristics were identified. Most interestingly, most tables are very wide and contain hundreds of columns. However, many attributes of such table are not used at all: 55 % of all columns are unused on average per company. This is due to the fact, that standard software needs to support many workflows in different industries and countries, however a single company never uses all of them. Further, in many columns NULL or default values are dominant, so the entropy (information containment) of these columns is very low (near zero).

But even the columns that are used by a specific company often have a low cardinality of values, i.e., there are very few distinct values. Often due to the fact that the data models the real world, and every company has only a limited number of products that can be sold, to a limited number of customers, by a limited number of employees and so on.

These characteristics facilitate the efficient use of compression techniques that we will introduce in Chap. 7, leading to lower memory consumption and better query performance.

3.7   Self Test Questions

1. **OLTP OLAP Separation Reasons**
   Why was OLAP separated from OLTP?

   (a) Due to performance problems
   (b) For archiving reasons; OLAP is more suitable for tape-archiving
   (c) Out of security concerns
   (d) Because some customers only wanted either OLTP or OLAP and did not want to pay for both

References

[Fre95]   C.D. French, "One size fits all" database architectures do not work for DSS. SIGMOD Rec. **24**(2), 449–450 (1995)

[KGZP10]   J. Krueger, M. Grund, A. Zeier, H. Plattner, Enterprise application-specific data management, in *EDOC*, pp. 131–140, 2010

[KKG⁺11]   J. Krueger, C. Kim, M. Grund, N. Satish, D. Schwalb, J. Chhugani, H. Plattner, P. Dubey, A. Zeier, Fast updates on read-optimized databases using multi-core CPUs. Proc. VLDB **5**(1), 61–72 (2011)

Chapter 4
Changes in Hardware

This chapter gives an overview of the foundations to understand how the changing hardware impacts software and application development. It is partly taken from [SKP12].

In the early 2000s multi-core architectures were introduced, starting a trend towards growing parallelism. Today, a typical enterprise computing board comprises eight CPUs and 10 to 16 cores per CPU, adding up to 80 to 128 cores per server. An address space of 64 bit ensures that the increasing amount of main memory can be addressed, with current servers supporting up to 6 TB. The maximum data throughput between a CPU and DRAM typically approximates 85 GB/s and it is possible to transfer more than 6 GB/s between servers via a typical InfiniBand FDR 4x link. Each of those systems offers a high level of parallel computing for a price of about $50,000.

Despite the introduction of massive parallelism, the disk totally dominated all thinking and performance optimizations not long ago. It was extremely slow, but necessary to store data. Compared to the speed development of CPUs, the development of disk performance could not keep up. This resulted in a complete distortion of the whole model of working with databases and large amounts of data. Today, the increasing size of main memory available in servers combined with its ever decreasing price initiate a shift from disk-based to main-memory based database management systems. These systems keep the primary copy of their data in main memory contrary to an architecture leveraging fast but small buffers and disk for persistence.

4.1 Memory Cells

In early computer systems, the clock rate of the CPU was the same as the frequency of the memory bus and register access was only slightly faster than memory access. CPU frequencies however heavily increased over the last years following Moore's

Law[1] [Moo65] while frequencies of memory buses and latencies of memory chips did not grow with this same speed. As a result memory accesses became increasingly expensive as more CPU cycles were wasted waiting for memory access. This development is not caused by the fact that fast memory cannot be built, but an economic decision. Memory that is as fast as current CPUs would be orders of magnitude more expensive and would require extensive physical space on the boards.

Random-access memory cells are divided into SRAM (Static Random Access Memory) and DRAM (Dynamic Random Access Memory) from an electrotechnical design perspective. SRAM cells are usually built out of six transistors per data bit (variants with only four exist but have major disadvantages [MSMH08]) and can store a stable state as long as power is supplied. Accessing the stored state requires raising the word access line and the state is immediately available for reading.

In contrast, DRAM cells can be constructed using a much simpler structure consisting of only one transistor and a capacitor per data bit. The state of the memory cell is stored in the capacitor while the transistor is only used to control access to the capacitor. This design has major advantages in required space and cost compared to SRAM, however, introduces a problem as capacitors discharge over time and while reading the state of the memory cell. To mitigate this problem modern systems refresh DRAM chips every 64 ms [CJDM01] and after every read of the cell in order to recharge the capacitor. During the refresh, no access to the state of the cell is possible. The charging and discharging of the capacitor requires time, therefore the current state cannot be detected immediately after the request to store the state thus limiting the speed of DRAM cells.

In a nutshell, SRAM is fast but requires more space whereas DRAM chips are slower but allow for a higher capacity due to their simpler structure and higher density. For more details about the two types of RAM and their electrotechnical design the interested reader is referred to [Dre07].

4.2   Memory Hierarchy

An underlying assumption of the memory hierarchy of modern computer systems is a principle known as *data locality* [HP03]. Temporal data locality indicates that data that is accessed is likely to be accessed again soon, whereas spatial data locality indicates that data stored adjacent in memory is likely to be accessed together. These principles are leveraged by using caches, combining the fast access to SRAM chips with the high capacity of DRAM chips.

Figure 4.1 gives an example of the memory and cache hierarchies on the Intel Ivy Bridge architecture. Small and fast caches close to the CPUs built of SRAM cells

---

[1]Moore's Law is the assumption that the number of transistors on integrated circuits doubles every 18–24 months. This assumption still holds true till today.

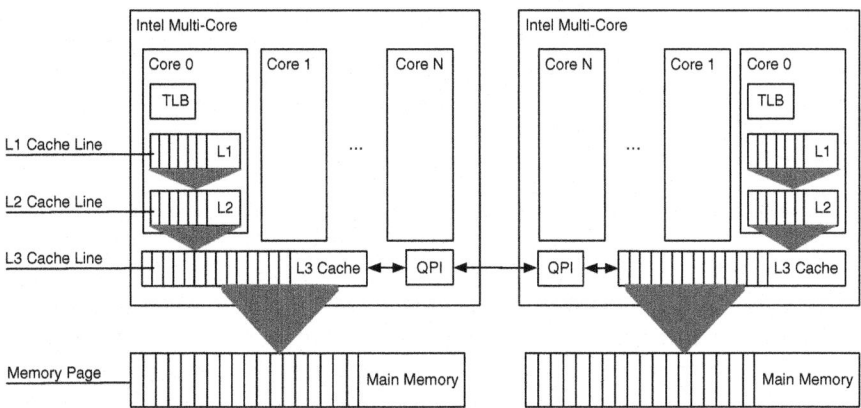

Fig. 4.1 Memory hierarchy on Intel Ivy Bridge-EX

act as an intermediate layer to buffer frequent accesses to the slower main memory built of DRAM cells. This concept is commonly implemented in the form of multi-level caches where data needs to traverse the different levels of caches characterized by increasing speed and decreasing size to the CPU registers to be used as input for instructions. The size and amount of those extremely small integer and floating point caches is usually very limited.

The cache hierarchy is usually implemented as a combination of private L1 (Level 1) and L2 (Level 2) caches combined with a shared L3 (Level 3) cache that all cores on one socket can access. To ensure low latencies and high bandwidth each socket houses an IMC (Integrated Memory Controller) to access the local part of the main memory. Remote memory access on the other hand needs to be performed over a QPI (Quick Path Interconnect) controller coordinating the access.

4.3 Cache Internals

Caches are organized in cache lines for management purposes. If the requested content cannot be located in any cache, it is loaded from main memory and transferred through the memory hierarchy to the registers. The smallest transferable unit between each level is one *cache line*. Caches where every cache line of level i is also present in level $i + 1$ are called *inclusive caches*. Otherwise the model is called *exclusive caches*. All Intel processors implement an inclusive cache model. This inclusive cache model is assumed for the rest of this text.

When requesting a cache line from the cache, the process of determining whether the requested line is already in the cache and locating where it is cached is crucial. In theory it is possible to implement fully associative caches, where each cache line can cache any memory location. In practice this is only feasible for very small caches

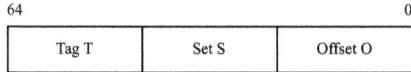

Fig. 4.2 Parts of a memory address

as a search over the complete cache becomes necessary when searching for a cache line. In order to minimize the region to be searched, the concept of an *n-way set associative cache* with associativity A_i divides a cache with C_i bytes in $C_i/B_i/A_i$ sets and restricts the number of cache lines which can contain a copy of a certain memory address to one set or A_i cache lines. When determining if a cache line is already present in the cache therefore only one set with A_i cache lines needs to be searched.

In order to determine if a requested address from main memory is already cached it is split into three parts as shown in Fig. 4.2. The first part is the offset O, whose size is determined by the cache line size of the cache. With a cache line size of 64 bytes, the lower 6 bits of the address would be used as the offset into the cache line. The number s of bits used to identify the cache set is determined by the cache size C_i, the cache line size B_i and the associativity A_i of the cache by $s = \log_2(C_i/B_i/A_i)$. The remaining 64-o-s bits of the address are used as a tag to identify the cached copy. When requesting an address from main memory the processor can therefore calculate S by masking the address and then search the respective cache set for the tag T. This can easily be done by comparing the tags of the A_i cache lines in the set in parallel.

4.4 Address Translation

The operating system provides a dedicated continuous address space for each process, containing an address range from 0 to 2^x. This has several advantages as the process can address the memory through virtual addresses and does not have to take the physical fragmentation into account. Memory protection mechanisms can additionally control the access to memory, restricting programs to access memory which was not allocated by them. Another advantage of virtual memory is the use of a paging mechanism which allows a process to use more memory than physically available by evacuating pages to secondary storage.

The continuous virtual address space of a process is divided into pages of size p. On most operating systems p is equal to 4 KB. Those virtual pages are mapped to physical memory. The mapping itself is saved in a so called page table that resides in main memory itself. When the process accesses a virtual memory address, the address is translated into a physical address by the operating system with help of the memory management unit inside the processor.

While not going into details of the translation and paging mechanisms, the address translation is usually done by a multi-level page table. The virtual address is split into multiple parts that are used as an index into the page directories resulting in a physical address and a respective offset. As the page table is kept in main memory, each translation of a virtual address into a physical address would require additional main memory accesses or cache accesses in case the page table is cached.

In order to speed up the translation process, the computed values are cached in the small and fast Translation Lookaside Buffer (TLB). When accessing a virtual address the respective tag for the memory page is calculated by masking the virtual address. The TLB is then used to look up the tag. In case the tag is found, the physical address can be directly retrieved from the cache. Otherwise a TLB miss occurs and the physical address needs to be calculated, which can be quite costly. Details about the address translation process, TLBs and paging structure caches for Intel 64 and IA-32 architectures can be found in [Int08].

The cost introduced with the address translation scales linearly with the width of the translated address [HP03, CJDM99], therefore making it hard or impossible to build large memories with very small latencies.

4.5 Prefetching

Modern processors try to predict which data will be accessed next, initiating loads before the data is actually accessed in order to reduce the incurring access latencies. Good prefetching can completely hide the latencies so that the data is already in the cache when accessed. If data is prefetched that is not accessed it can, however, also evict data that would be accessed later and therefore induce additional cache misses. Processors support software and hardware prefetching. Software prefetching can be seen as a hint to the processor, indicating which addresses are accessed next. Hardware prefetching automatically recognizes access patterns by utilizing different prefetching strategies. The Intel Ivy Bridge architecture contains two second level cache prefetchers—the L2 streamer and data prefetch logic (DPL) [Int11].

4.6 Memory Hierarchy and Latency Numbers

The memory hierarchy can be seen as a pyramid of storage mediums. The slower a medium, the cheaper it gets. This also means that the storage size on the slower levels increases with its lower price. The hierarchy levels of modern hardware are outlined in Fig. 4.3.

At the very bottom, the cheapest and biggest medium is the hard disk. It replaces magnetic tapes as the slowest storage medium.

Located on the next level, Flash is significantly faster than a traditional disk, but still used like one from a software perspective because of its persistence and usage characteristics. This means that the same block oriented input and output methods developed more than 20 years ago for disks are still used for flash. In order to fully utilize the speed of flash based storage the interfaces and drivers need to be adapted accordingly.

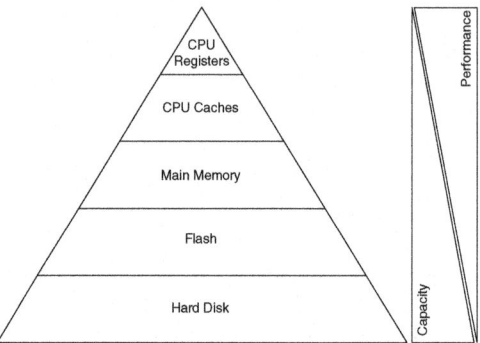

Fig. 4.3 Conceptual view of the memory hierarchy

Table 4.1 Latency numbers

Action	Time (ns)	Time
L1 cache reference (cached data word)	0.5	
Branch mispredict	5	
L2 cache reference	7	
Mutex lock/unlock	25	
Main memory reference	100	0.1 μs
Send 2,000 byte over 1 Gb/s network	20,000	20 μs
SSD random read	150,000	150 μs
Read 1 MB sequentially from memory	250,000	250 μs
Disk seek	10,000,000	10 ms
Send packet CA to Netherlands to CA	150,000,000	150 ms

The directly accessible main memory is located between Flash and the CPU caches L3, L2 and L1. The CPU registers, where data needs to be located to be used in actual calculations, form the top of the memory hierarchy. As every operation takes place inside the CPU and in turn the data has to be in the registers, there are usually four layers that are only used for transporting information when accessing data from disk.

Table 4.1 gives an overview of some of the latencies related to the memory hierarchy. Latency is the time delay that has to be taken into account for the system to load data from the respective storage medium into the CPU registers. The L1 cache latency is 0.5 ns. In contrast, accessing a main memory reference takes 100 ns and a simple disk seek accounts for a 10 ms delay.

In the end, there is nothing special about the main-memory based approach for database management systems. All computing ever done was in memory as actual calculations can only take place in the CPU. The performance of an application using large amounts of data is usually determined by how fast data can be transferred through the memory hierarchy to the CPU registers. To estimate the runtime of a typical database operator algorithm, it is therefore possible to roughly calculate the amount of data that needs to be transferred to the CPU. The delay caused by the transportation is decreased when data is loaded from main memory instead of fetching it from disk. This thereby improves the overall system performance.

One of the simplest operations a CPU can perform is a comparison, used for instance when filtering for an attribute. With an assumed scan-speed of 4 MB/ms for this operation on one core the theoretical scan-speed on a 15-core CPU (such as Intels Ivy Bridge EX) is 60 GB/s. On a typical 8-socket node this adds up to 480 GB/s for a filtering operation.

Considering the availability of large multi-node systems, having 10 nodes and 120 cores per node with the data distributed evenly across nodes, it's hard to find a use case in enterprise computing where an algorithm should require more than 1 s to complete.

4.7 Non-Uniform Memory Access

As the development in modern computer systems is shifting from multi-core to many-core systems and the size of the main memory continues to increase, using a Front Side Bus (FSB) architecture with a Uniform Memory Access (UMA) became a performance bottleneck and introduced heavy challenges in hardware design to connect all cores and memory.

Non-Uniform Memory Access (NUMA) attempts to solve this problem by introducing local memory locations with cheap access for the respective processing units. Figure 4.4 shows a rough comparison between a Uniform Memory Access (UMA) and a Non-Uniform Memory Access (NUMA) system architecture.

A UMA system is characterized by a deterministic access time for an arbitrary memory address independent of which processor makes the request as every memory chip is accessed through a central memory bus as shown in Fig. 4.4a. For NUMA systems on the other hand the access time depends on the memory location relative to the processor. Local (adjacent) memory can be accessed faster than non-local (adjacent to another processor) memory or shared memory (shared amongst processors) as shown in Fig. 4.4b.

NUMA systems are additionally classified into cache-coherent NUMA (ccNUMA) and non cache-coherent NUMA systems. ccNUMA systems provide each CPU the same view to the complete memory and enforce coherency using a protocol implemented in hardware. Non cache-coherent NUMA systems require software layers to handle memory conflicts accordingly. Although non ccNUMA hardware is easier and cheaper to build, most of todays available standard hardware provides ccNUMA, since non ccNUMA hardware is more difficult to program.

To fully utilise the potentials of NUMA, applications have to be made aware of the different memory locations and should primarily load data from the local memory of a processor. Memory-bound applications may suffer a degradation of up to 25 % of their maximum performance if non-local memory is accessed instead of local memory.

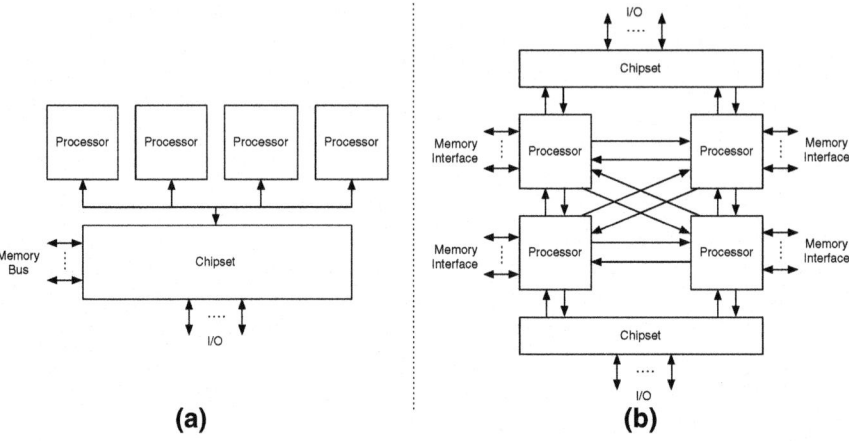

Fig. 4.4 (**a**) Shared FSB; (**b**) Intel Quick Path Interconnect [Int09]

4.8 Scaling Main Memory Systems

Figure 4.5 shows an example setup for scaling main-memory based database management systems using multiple nodes (scale-out). Each node comprises 8 CPUs with 15 cores adding up to 120 cores on one node and 480 cores on the 4 nodes shown. With every node containing 2 TB of DRAM the system can store a total of 8 TB of data completely in memory. Using SSDs or traditional disks for persistence such as logging, archiving and for emergency reconstruction of data, the system can easily reload the data to memory in case of a planned or unplanned restart of one or multiple nodes.

The network connecting the nodes is continuously increasing speed as well. In the example shown in Fig. 4.5 the nodes are connected via 10 Gbit/s Ethernet (GbE). With systems leveraging 56 Gbit/s InfiniBand FDR 4x connections already on the market and manufacturers talking about speeds as high as several 100 Gbit/s the viability of vertical parallelism (scale out) will increase even more. In the context of main-memory based database management systems this trend will mainly improve join queries for which data needs to be fetched from multiple nodes.

4.9 Remote Direct Memory Access

The use of shared memory to directly access memory on remote nodes is increasingly becoming an alternative to traditional network-based communication. For nodes connected via an InfiniBand link it is possible to establish a shared memory region to automatically access data on different nodes without explicitly requesting the data from the respective node. This direct access without the need for shipping and processing on the remote node is a very light-weight way of vertically scaling

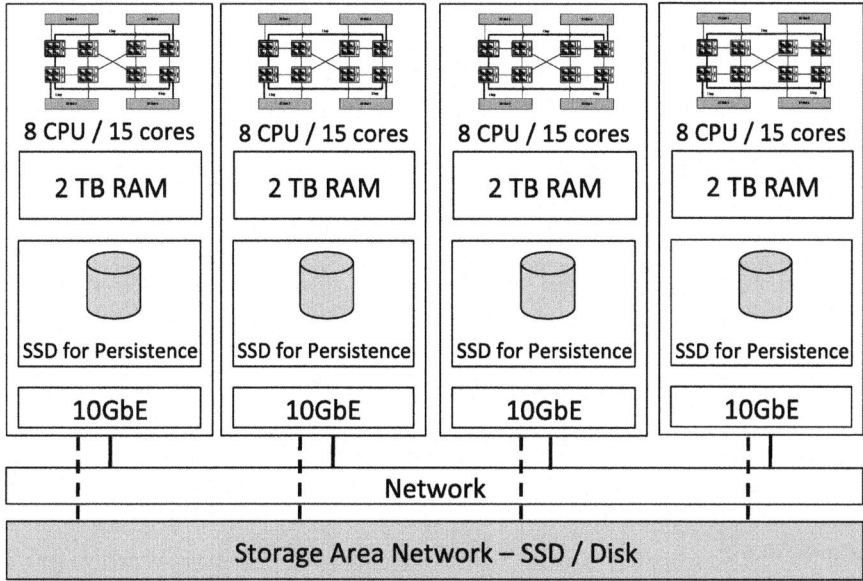

Fig. 4.5 A system consisting of multiple blades

computational capacity. Research done at Stanford University in cooperation with the HPI using a large-scale RAM cluster is showing very promising results thus offering direct access to a seemingly unlimited amount of memory with very little overhead.

4.10 Self Test Questions

1. **Speed per Core**
   What is the speed of a single core when processing a simple scan operation (under optimal conditions)?

   (a) 4 GB/ms/core
   (b) 4 MB/ms/core
   (c) 4 MB/s/core
   (d) 400 MB/s/core

2. **Latency of Hard Disk and Main Memory**
   Which statement concerning latency is wrong?

   (a) The latency of main memory is about 100 ns
   (b) A disk seek takes an average of 0.5 ms
   (c) Accessing main memory is about 100,000 times faster than a disk seek
   (d) Ten milliseconds is a good estimation for a disk seek

References

[CJDM99]   V. Cuppu, B. Jacob, B. Davis, T. Mudge, A performance comparison of contemporary DRAM architectures, in *Proceedings of the 26th Annual International Symposium on Computer Architecture*, 1999

[CJDM01]   V. Cuppu, B. Jacob, B. Davis, T. Mudge, High-performance DRAMs in workstation environments. IEEE Trans. Comput. **50**(11), 1133–1153 (2001)

[Dre07]   U. Drepper, What every programmer should know about memory (2007), http://people.redhat.com/drepper/cpumemory.pdf

[HP03]   J. Hennessy, D. Patterson, *Computer Architecture: A Quantitative Approach* (Morgan Kaufmann, San Francisco, 2003)

[Int08]   Intel Inc., *TLBs, Paging-Structure Caches, and Their Invalidation*, 2008

[Int09]   Intel Inc., *An Introduction to the Intel QuickPath Interconnect*, 2009

[Int11]   Intel Inc., *Intel 64 and IA-32 Architectures Optimization Reference Manual*, 2011

[MSMH08]   A.A. Mazreah, M.R. Sahebi, M.T. Manzuri, S.J. Hosseini, A novel zero-aware four-transistor SRAM cell for high density and low power cache application, in *International Conference on Advanced Computer Theory and Engineering, 2008, ICACTE '08*, 2008, pp. 571–575

[Moo65]   G. Moore, Cramming more components onto integrated circuits. Electronics **38**, 114 ff. (1965)

[SKP12]   D. Schwalb, J. Krueger, H. Plattner, Cache conscious column organization in in-memory column stores. Technical Report 60, Hasso-Plattner-Institute, December 2012

Chapter 5
A Blueprint of SanssouciDB

SanssouciDB is a prototypical database system for unified analytical and transactional processing. The concepts of SanssouciDB build on prototypes developed at the HPI and an existing SAP database system. SanssouciDB is an SQL database and it contains similar components as other databases such as a query builder, a plan executer, meta data, a transaction manager, etc.

5.1 Data Storage in Main Memory

In contrast to traditional database management systems, the primary persistence of SanssouciDB is main memory. Yet logging and recovery still require disks as non-volatile data storage to ensure data consistency in case of failures. All operators, e.g., find, join, or aggregation can anticipate that data resides in main memory. Thus, operators are implemented differently moving the focus from optimizing for disk access towards optimizing for main memory access and CPU utilization (see Chap. 4).

This apparently subtle difference of moving the primary persistence has a vast impact on performance even when disk-based databases are completely memory resistent. Ailamaki et al. invested such fully cached disk-based databases and found that a large portion of query execution is spent for memory and resource stalls [ADHW99]. Those stalls are mainly caused by in-page data placements that do not utilize the CPU caches properly. In many cases, the actual computation accounts for less than 40 % of the execution time. Besides, Harizopoulos et al. found that the buffer management of disk-based databases alone contributes 31 % to the overall instruction count [HAMS08]

Consequently, the reason for the performance advantages of in-memory over disk-based databases derives from optimized data structures and algorithms avoiding memory and resource stalls together with the removal of additional indirections.

H. Plattner, *A Course in In-Memory Data Management*,
DOI 10.1007/978-3-642-55270-0_5, © Springer-Verlag Berlin Heidelberg 2014

5.2   Column-Orientation

Another concept used in SanssouciDB was invented more than 2 decades ago, that is, storing data column-wise [CK85] instead of row-wise. In column-orientation, complete columns are stored in adjacent blocks. This can be contrasted with row-oriented storage where complete tuples (rows) are stored in adjacent blocks. Column-oriented storage, in contrast to row-oriented storage, is well suited for reading consecutive entries from a single column. This can be useful for aggregation and column scans. More details on column-orientation and its differences to row-orientation can be found in Chap. 8. To minimize the amount of data that needs to be transferred between storage and processor, SanssouciDB uses several different data compression techniques, which will be discussed in Chap. 7.

5.3   Implications of Column-Orientation

Column-oriented storage has become widespread in database systems specifically developed for OLAP, as the advantage of column-oriented storage is clear in case of quasi-sequential scanning of single attributes and set processing thereof. If not all fields of a table are queried, column-orientation can be exploited as well in transactional processing (avoiding "SELECT *"). An analysis of enterprise applications showed that there is actually no application that uses all fields of a given tuple. For example, in dunning only 17 attributes are necessary out of a table that contains 300 attributes. If only the 17 needed attributes are queried instead of the full tuple representation of all 300 attributes, an instant advantage of factor 8–20 for data to be scanned can be achieved.

As disk is not the bottleneck any longer, but access to main memory has to be considered, an important aspect is to work on a minimal set of data. So far, application programmers were fond of "SELECT *" statements. The difference in runtime between selecting specific fields or all fields in row-oriented storage is often insignificant and in case changes to an application need more fields, the data was already there (which besides is a weak argument for using SELECT * and retrieving unnecessary data). However, in case of column-orientation, the penalty for "SELECT *" statements grows with table width. Especially if tables are growing in width during productive usage, actual runtimes of applications cannot be anticipated during programming.

With the column-store approach, the number of indices can be significantly reduced. In a column store, every attribute can be used as an index. Because all data is available in memory and the data of a column is stored consecutively, the scanning speed is high enough that a full sequential scan of an attribute is sufficient in most cases. If this is not fast enough, dedicated indices can still be used in addition for further speedup.

Storing data in columns instead of rows is challenging for workloads with many data modifying operations. Therefore, the concept of a differential buffer was

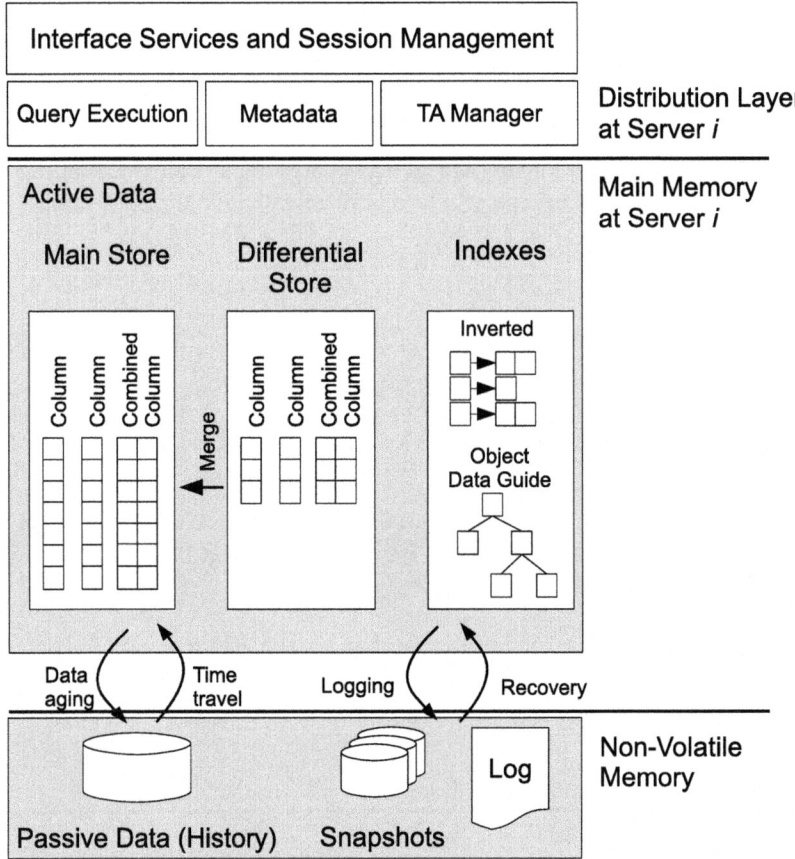

Fig. 5.1 Schematic architecture of SanssouciDB

introduced, where new entries are written to a differential buffer first. In contrast to the main store, the differential buffer is optimized for inserts. At a later point in time and depending on thresholds, e.g. the frequency of changes and new entries, the data in the differential buffer is merged into the main store. More details about the differential buffer and the merge process will be provided later in Chaps. 25 and 27.

5.4 Architecture Overview

The architecture shown in Fig. 5.1 grants an overview of the components of SanssouciDB.

SanssouciDB is split in three different logical layers fulfilling specific tasks inside the database system. The "Distribution Layer" handles the communication to applications, creates query execution plans, stores meta data and contains the logic for database transactions. Inside the main memory of a specific machine the main working set of SanssouciDB is located. That working set is accessed during query execution and is stored either in row, column or hybrid-oriented data layout, depending on the specific type of queries sent to the database tables. The non-volatile memory is used for logging and recovery purposes, as well as for data aging and time travel. All those concepts will be described in the subsequent sections.

References

[ADHW99]   A. Ailamaki, D.J. DeWitt, M.D. Hill, D.A. Wood, DBMSs on a modern processor: where does time go? in *VLDB*, ed. by M.P. Atkinson, M.E. Orlowska, P. Valduriez, S.B. Zdonik, M.L. Brodie (Morgan Kaufmann, San Francisco, 1999), pp. 266–277

[CK85]   G.P. Copeland, S.N. Khoshafian, A decomposition storage model. SIGMOD Rec. **14**(4), 268–279 (1985)

[HAMS08]   S. Harizopoulos, D.J. Abadi, S. Madden, M. Stonebraker, Oltp through the looking glass, and what we found there, in *SIGMOD Conference*, ed. by J.T.-L. Wang, (ACM, 2008), pp. 981–992

Part II
Foundations of Database Storage Techniques

Chapter 6
Dictionary Encoding

Since memory is the new bottleneck, it is required to minimize access to it. Accessing a smaller number of columns by only querying required attributes can do this on the one hand. On the other hand, decreasing the number of bits used for data representation can reduce both memory consumption and transfer times.

Dictionary encoding builds the basis for several other compression techniques (see Chap. 7) that might be applied on top of the encoded columns. The main effect of dictionary encoding is that long values, such as texts, are represented as short integer values.

Dictionary encoding is relatively simple. This does not only mean that it is easy to understand, but also that it is easy to implement and does not have to rely on complex multilevel procedures, which would limit or lessen the performance gains. First, we will use the example presented in Fig. 6.1 to explain the general algorithm that translates original values to integers.

Dictionary encoding is applied to each column of a table separately. In the example, every distinct value in the first name column "fname" is replaced by a distinct integer value. The position of a text value (e.g. Mary) in the dictionary is the representing number for that text (here: "24" for Mary). Until now, we have not saved any storage space. The benefits come to effect with values appearing more than once in a column. In our small example, the value "John" can be found twice in the column "fname", namely on position 39 and 42. Using dictionary encoding, the long text value (we assume 49 Byte per entry in the first name column) is represented by a short integer value (23 bit are needed to encode the 5 million different first names we assume to exist in the world). The more often identical values appear, the better dictionary encoding can compress a column. As we noted in Sect. 3.6, enterprise data has low entropy. Therefore, dictionary encoding is well suited and yields a good compression ratio in such scenarios. In the following, we calculate the possible savings for the first name and gender columns of our world-population example.

H. Plattner, *A Course in In-Memory Data Management*,
DOI 10.1007/978-3-642-55270-0_6, © Springer-Verlag Berlin Heidelberg 2014

Column "fname"			Dictionary for "fname"			Attribute Vector for "fname"	
recID	fname		valueID	Value		position	valueID
...	...		...	...		...	...
39	John		23	John		39	23
40	Mary		24	Mary		40	24
41	Jane		25	Jane		41	25
42	John		26	Peter		42	23
43	Peter		...	...		43	26
...	...					...	...

Fig. 6.1 Dictionary encoding example

6.1   Compression Example

Given the world population table with 8 billion rows and 200 Byte per row:

Attribute	# of distinct values	Size (Byte)
First name	5 million	49
Last name	8 million	50
Gender	2	1
Country	200	49
City	1 million	49
Birthday	40,000	2
	Sum	200

The complete amount of data is:

$$8 \ billion \ rows \cdot 200 \ Byte \ per \ row = 1.6 \ TB$$

Each column is split into a dictionary and an attribute vector. Each dictionary stores all distinct values present within the column. The valueID of each value is implicitly given by the value's position in the dictionary and, thus, does not need to be stored explicitly.

In a dictionary-encoded column, the attribute vectors now only store valueIDs, which correspond to the valueIDs in the dictionary. The recordID (row number) is stored implicitly via the position of an entry in the attribute vector. To sum up, via dictionary encoding, all information can be stored as integers instead of other, usually larger, data types.

6.1.1   Dictionary Encoding Example: First Names

How many bits are required to represent all 5 million distinct values of the first name column "fname"?

$$\lceil log_2(5{,}000{,}000) \rceil = 23$$

Therefore, 23 bits are enough to represent all distinct values for the required column. Instead of using

$$8\ billion \cdot 49\ Byte\ =\ 392\ billion\ Byte\ =\ 365.1\ GB$$

for the first name column, the attribute vector itself can be reduced to the size of

$$8\ billion\ \cdot\ 23\ bit\ =\ 184\ billion\ bit\ =\ 23\ billion\ Byte\ =\ 21.4\ GB$$

and an additional dictionary is introduced, which needs

$$49\ Byte\ \cdot\ 5\ million\ =\ 245\ million\ Byte\ =\ 0.23\ GB.$$

The achieved compression factor can be calculated as follows:

$$\frac{uncompressed\ size}{compressed\ size} = \frac{365.1\ GB}{21.4\ GB + 0.23\ GB} \approx 17$$

That means we reduced the column size by a factor of 17 and the resulting data structures only consume about 6 % of the initial amount of main memory.

6.1.2 Dictionary Encoding Example: Gender

Now, let us look at the gender column. It only has 2 distinct values. In order to represent the gender ("m" or "f") without compression for each value, 1 Byte is required. So, without compression, the data amounts to:

$$8\ billion \cdot 1\ Byte = 7.45\ GB$$

If compression is used, then 1 bit is enough to represent the same information. The attribute vector requires:

$$8\ billion \cdot 1\ bit = 8\ billion\ bit = 0.93\ GB$$

of space. The dictionary need an additional:

$$2 \cdot 1\ Byte = 2\ Byte$$

This amounts to a compression factor of:

$$\frac{uncompressed\ size}{compressed\ size} = \frac{7.45\ GB}{0.93\ GB + 2\ Byte} \approx 8$$

The compression rate depends on the size of the initial data type as well as on the column's entropy, which is determined by two cardinalities:

- Column cardinality: the number of distinct values in a column
- Table cardinality: the total number of rows in the table or column

Entropy is a measure which expresses how much information is contained within a column. It is calculated as follows:

$$entropy = \frac{column\ cardinality}{table\ cardinality}$$

The smaller the entropy of the column, the better the achievable compression rate.

6.2   Sorted Dictionaries

The benefits of dictionary encoding increase if sorting is applied to the dictionary. Retrieving a value from a sorted dictionary speeds up the lookup process from $O(n)$, which means a full scan through the dictionary, to $O(log(n))$, because values in the dictionary can be found using binary search. But this optimization comes at a cost. The dictionary has to be re-sorted every time a new value that does not belong at the end of the sorted sequence is added to the dictionary. In this case, the positions of already present values that are behind the inserted value have to be pushed back by one position. While sorting the dictionary is not that costly, updating the corresponding attribute vector is. In our example, about 8 billion values would have to be checked or updated if, for example, a new first name is added to the dictionary.

6.3   Operations on Encoded Values

The first and most important effect of dictionary encoding is that all operations concerning the table data are now done via attribute vectors, which solely consist of integers. This causes an implicit speedup of all operations, since a CPU is designed to perform operations on numbers, not on characters. When explaining dictionary encoding, one question is asked quite often: "But isn't the process of looking up all values via an additional data structure more costly than the actual savings? We understand the benefits concerning main memory, but what about the processor?"— This, of course, is an appropriate and important question. The processor has to take additional load. But given the fact that our bottlenecks are memory and bandwidth, a slight shift of workload in the direction of the processor is not only acceptable but also welcome. Secondly, the impact of retrieving the actual values for the encoded columns is rather small. When selecting tuples, only the corresponding values defined in the where clause of the query have to be looked up in the dictionary in order to perform the column scan. Generally, result sets should be rather small compared to the total table size, so the lookup of all other selected columns, which is necessary to materialize the query result, is only problematic for queries with a high selectivity (i.e., ones that return a large amount of the overall records present within the table). Carefully written queries avoid returning huge amounts of records and also only select those columns that are really needed within the application. These simple optimizations not only save bandwidth, but also further reduce the number of necessary lookups. Finally, several operations such as COUNT or NOT NULL can be performed without retrieving the real values at all and, thus, can be executed on dictionary encoded columns without additional performance overhead.

6.4 Self Test Questions

1. **Lossless Compression**
   For a column with few distinct values, how can dictionary encoding significantly reduce the required amount of memory without any loss of information?

   (a) By mapping values to integers using the smallest number of bits possible to represent the given number of distinct values
   (b) By converting everything into full text values. This allows for better compression techniques, because all values share the same data format.
   (c) By saving only every second value
   (d) By saving consecutive occurrences of the same value only once

2. **Compression Factor on Whole Table**
   Given a population table (50 millions rows) with the following columns:

   - name (49 bytes, 20,000 distinct values)
   - surname (49 bytes, 100,000 distinct values)
   - age (1 byte, 128 distinct values)
   - gender (1 byte, 2 distinct values)

   What is the compression factor (uncompressed size/compressed size) when applying dictionary encoding?

   (a) ≈ 20
   (b) ≈ 90
   (c) ≈ 10
   (d) ≈ 5

3. **Information in the Dictionary**
   What information is saved in a dictionary in the context of dictionary encoding?

   (a) Cardinality of a value
   (b) All distinct values
   (c) Hash of a value of all distinct values
   (d) Size of a value in bytes

4. Advantages through Dictionary Encoding

What is an advantage of dictionary encoding?

(a) Sequentially writing data to the database is sped up
(b) Aggregate functions are sped up
(c) Raw data transfer speed between application and database server is increased
(d) INSERT operations are simplified

5. Entropy

What is entropy?

(a) Entropy limits the amount of entries that can be inserted into a database. System specifications greatly affect this key indicator.
(b) Entropy represents the amount of information in a given dataset. It can be calculated as the number of distinct values in a column (column cardinality) divided by the number of rows of the table (table cardinality).
(c) Entropy determines tuple lifetime. It is calculated as the number of duplicates divided by the number of distinct values in a column (column cardinality).
(d) Entropy limits the attribute sizes. It is calculated as the size of a value in bits divided by number of distinct values in a column the number of distinct values in a column (column cardinality).

Chapter 7
Compression

As discussed in Chap. 5, SanssouciDB is a database designed to run transactional and analytical workloads in enterprise computing. The underlying data set can easily reach a size of several terabytes in large companies. Although memory capacities of commodity servers are growing, it is still expensive to process those huge data sets entirely in main memory. Therefore, SanssouciDB and several other modern in-memory storage engines use compression techniques on top of the initial dictionary encoding to decrease the total memory requirements. Columnar storage of data is well suited for compression, as data of the same type and domain is stored consecutively and can thus be processed efficiently.

Another advantage of compression is that it reduces the amount of data that needs to be transferred between main memory and CPUs, thereby increasing the performance of query execution. We discuss this in more detail in Chap. 16 on materialization strategies.

This chapter introduces several lightweight compression techniques, which provide a good trade-off between compression rate and additional processing overhead for encoding and decoding. There are also a large number of so-called heavyweight compression techniques. They achieve much higher compression rates, but encoding and decoding is prohibitively expensive for their usage in the context of enterprise applications. An in-depth discussion of many compression techniques can be found in [LSFZ10, AMF06].

7.1 Prefix Encoding

In real-world databases, we often find the case that a column contains one predominant value while the remaining values appear only seldom. Under this circumstance, we would store the same value very often in an uncompressed format. Prefix encoding is a simple way to handle this situation more efficiently. To apply prefix encoding, the data sets need to be sorted by the column with the predominant value and the attribute vector has to start with the predominant value.

H. Plattner, *A Course in In-Memory Data Management*,
DOI 10.1007/978-3-642-55270-0_7, © Springer-Verlag Berlin Heidelberg 2014

(a)   **(b)**

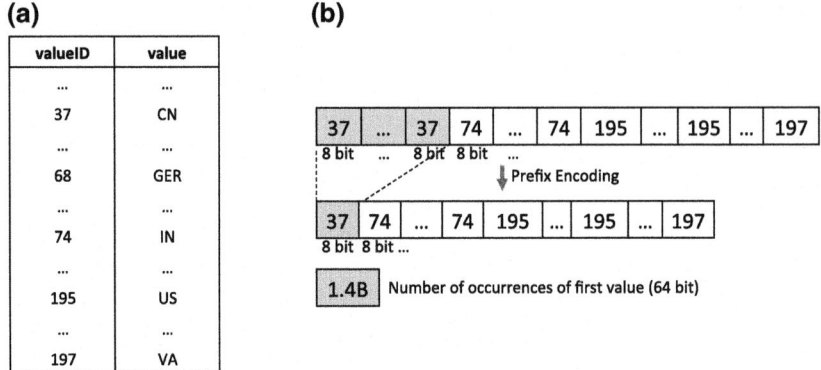

Fig. 7.1 Prefix encoding example. (**a**) Dictionary, (**b**) dictionary-encoded attribute vector (*top*) and prefix-encoded dictionary-encoded attribute vector (*bottom*)

To compress the column, the predominant value should not be stored explicitly every time it occurs. This is achieved by saving the number of occurrences of the predominant value and one instance of the value itself in the attribute vector. Thus, a prefix-encoded attribute vector contains the following information:

- number of occurrences of the predominant value
- valueID of the predominant value from the dictionary
- valueIDs of the remaining values

7.1.1   Example

Given is the attribute vector of the country column from the world population table, which is sorted by population of countries in descending order. Thus, the 1.4 billion Chinese citizens are listed first, then Indian citizens and so on. The valueID for China, which is situated at position 37 in the dictionary (see Fig. 7.1a), is stored 1.4 billion times at the beginning of the attribute vector in uncompressed format. In compressed format, the valueID 37 will be written only once, followed by the remaining valueIDs for the other countries as before. The number of occurrences "1.4 billion" for China will be stored explicitly. Figure 7.1b depicts the uncompressed and compressed attribute vectors for this example.

The following calculation illustrates the compression rate. First of all the number of bits required to store all 200 countries is calculated as $\lceil log_2(200) \rceil$ which results in 8 bit.

Without compression the attribute vector stores the 8 bit for each valueID 8 billion times:

$$8 \; billion \cdot 8 \; bit = 8 \; billion \; Byte = 7.45 \; GB$$

If the country column is prefix-encoded, the valueID for China is stored only once in 8 bit instead of 1.4 billion times 8 bit. An additional 64 bit integer field is added to store the number of occurrences. Consequently, instead of storing 1.4 billion times 8 bit, only 64 bit + 8 bit = 72 bit are really necessary. The complete storage space for the compressed attribute vector is now:

$$(8\ billion - 1.4\ billion) \cdot 8\ bit + 64\ bit + 8\ bit = 6.15\ GB$$

Thus, in our example 1.3 GB (17 %) of storage space is saved. Another advantage of prefix encoding is the possibility to directly access a position, as the location in the compressed attribute vector can be calculated. For example, to find all male Chinese the database engine can determine that only tuples with row numbers from 1 to 1.4 billion should be considered and then filtered by the gender value.

Although we see that we have reduced the required amount of main memory, it is evident that we still store redundant information for all other countries. Therefore, we introduce run-length encoding in the next section.

7.2 Run-Length Encoding

Run-length encoding is a compression technique that works best if the attribute vector consists of few distinct values with a large number of occurrences. For maximum compression rates, the column needs to be sorted, so that all the same values are located together. In run-length encoding, value sequences with the same value are replaced with a single instance of the value and

- either its number of occurrences or
- its starting position as offsets.

Figure 7.2 provides an example of run-length encoding using the starting positions as offsets. Storing the starting position speeds up access, as we can find the correct offset via binary search.

7.2.1 Example

Applied to our example of the country column sorted by population, instead of storing all 8 billion values (7.45 GB), we store two vectors:

- one with all distinct values: 200 times 8 bit
- the other with starting positions: 200 times 33 bit with 33 bit necessary to store the offsets up to 8 billion ($\lceil log_2(8\ billion) \rceil = 33$ bit). An additional 33 bit field at the end of this vector stores the number of occurrences for the last value.

(a)   **(b)**

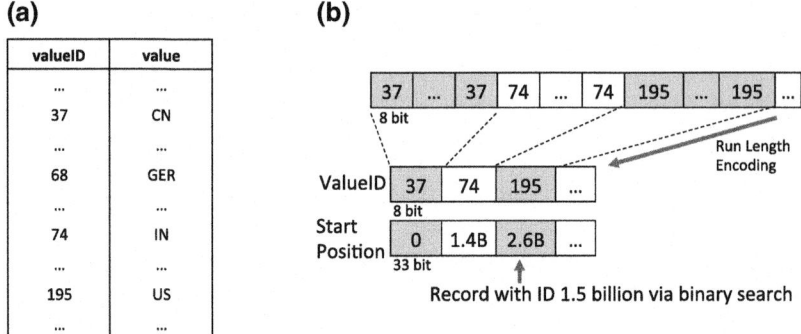

valueID	value
...	...
37	CN
...	...
68	GER
...	...
74	IN
...	...
195	US
...	...

Fig. 7.2 Run-length encoding example. (**a**) Dictionary, (**b**) dictionary-encoded attribute vector (*top*) and compressed dictionary-encoded attribute vector (*bottom*)

Hence, the size of the attribute vector can be significantly reduced to approximately 1 kB without any loss of information:

$$200 \cdot (33\ bit + 8\ bit) + 33\ bit \approx 1\ kB$$

7.3 Cluster Encoding

Cluster encoding works on equal-sized blocks of a column. The attribute vector is partitioned into N blocks of fixed size (typically 1,024 elements). If a cluster contains only a single value, it is replaced by a single occurrence of this value. Otherwise, the cluster remains uncompressed. An additional bit vector of length N indicates which blocks have been replaced by a single value (1 if replaced, 0 otherwise). Figure 7.3 depicts an example for cluster encoding with the uncompressed attribute vector on the top and the compressed attribute vector on the bottom. Here, the blocks only contain four elements for simplicity.

7.3.1 Example

Given is the city column (1 million different cities) from the world population table. The whole table is sorted cascadingly by country and city. Hence, cities, of the same country, are stored next to each other. Consequently, the occurrences of the same city values are stored next to each other, as well. Twenty bit are needed to represent 1 million city valueIDs ($\lceil log_2(1\ million)\rceil = 20$ bit). Without compression, the city attribute vector requires 18.6 GB (8 billion times 20 bit).

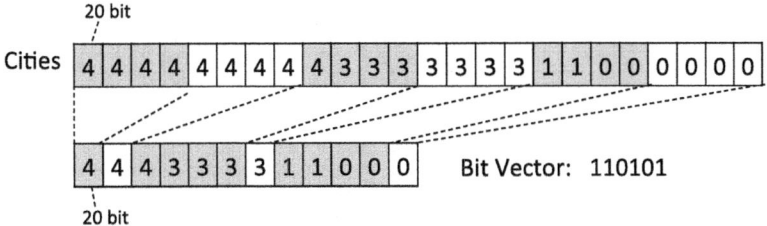

Fig. 7.3 Cluster encoding example with a block size of 4

Now, we compute the size of the compressed attribute vector illustrated in Fig. 7.3. With a cluster size of 1,024 elements the number of blocks is 7.8 million ($\lceil \frac{8\ billion\ rows}{1024\ elements\ per\ block} \rceil$). In the worst case every city leads to 1 incompressible block. Thus, the size of the compressed attribute vector is computed from the following sizes:

$$incompressible\ blocks + compressible\ blocks + bit\ vector$$

$$= 1\ million \cdot 1024 \cdot 20\ bit + (7.8 - 1)\ million \cdot 20\ bit + 7.8\ million \cdot 1\ bit$$

$$= 2,441\ MB + 16\ MB + 1\ MB$$

$$\approx 2.4\ GB$$

With a resulting size of 2.4 *GB*, a compression rate of 87 % (16.2 GB less space required) can be achieved in our example.

Cluster encoding does not support direct access to records. The position of a record needs to be computed via the bit vector. As an example, consider the query that counts how many men and women live in Berlin (for simplicity, we assume that only one city with the name "Berlin" exists and the table is sorted by city).

SELECT gender, **COUNT**(gender)
FROM world_population
WHERE city = 'Berlin'
GROUP BY gender

To find the recordIDs for the result set, we look up the valueID for "Berlin" in the dictionary. In our example, illustrated in Fig. 7.4, this valueID is 3. Then, we scan the cluster-encoded city attribute vector for the first appearance of valueID 3. While scanning the cluster-encoded vector, we need to maintain the corresponding position in the bit vector, as each position in the vector is mapped to either one value (if the cluster is compressed) or four values (if the cluster is uncompressed) of the cluster-encoded city attribute vector. In Fig. 7.4, this is illustrated by stretching the bit vector to the corresponding value or values of the cluster-encoded attribute vector. After the position is found, a bit vector lookup is needed to check whether the block(s) containing this valueID are compressed or not to determine the recordID range

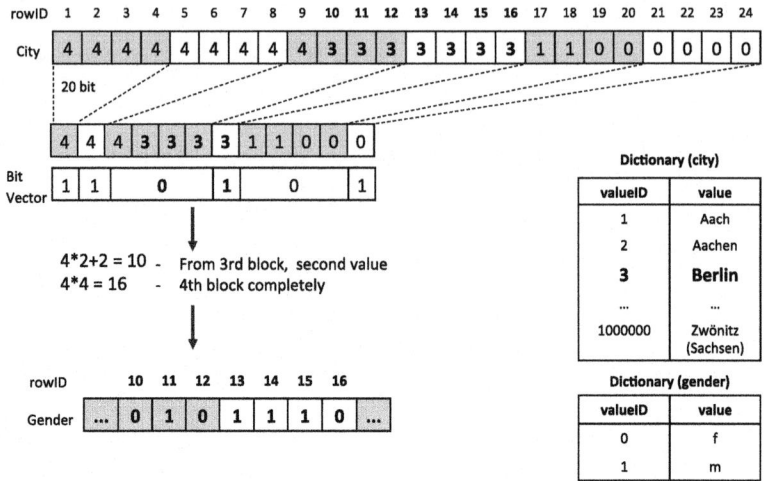

Fig. 7.4 Cluster encoding example: no direct access possible

containing the value "Berlin". In our example, the first block containing "Berlin" is uncompressed and the second one is compressed. Thus, we need to analyze the first uncompressed block to find the first occurrence of valueID 3, which is the second position, and can calculate the range of recordIDs with valueID 3, in our example 10 to 16. Having determined the recordIDs that match the desired city attribute, we can use these recordID to access the corresponding gender records and aggregate according to the gender values.

7.4 Indirect Encoding

Similar to cluster encoding, indirect encoding operates on blocks of data with N elements (typically 1,024). Indirect Encoding can be applied efficiently if data blocks hold a few distinct values. It is often the case if a table is sorted by another column and a correlation between these two columns exists (e.g., name column if table is sorted by countries).

Besides a global dictionary used by dictionary encoding in general, additional local dictionaries are introduced for those blocks that contain only a few distinct values. A local dictionary for a block contains all (and only those) distinct values that appear in this specific block. Thus, mapping the larger global valueIDs to even smaller local valueIDs can save space. Direct access is still possible, however, an indirection is introduced because of the local dictionary. Figure 7.5 depicts an example for indirect encoding with a block size of 1,024 elements. The upper part

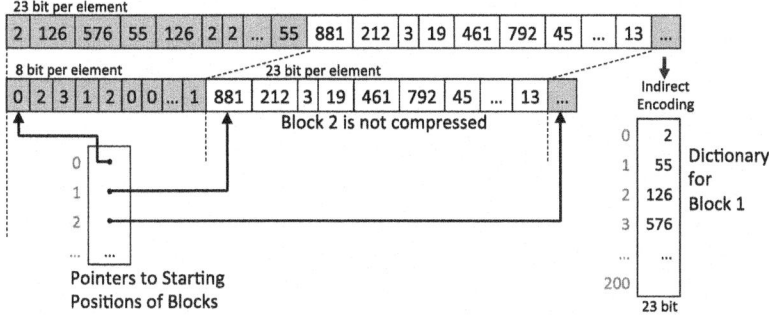

Fig. 7.5 Indirect encoding example

shows the dictionary-encoded attribute vector, the lower part shows the compressed vector, a local dictionary and the used pointer structure to directly access the starting positions of the respective blocks. This is necessary, because the blocks may use different amounts of bits to represent their values, as shown in the example. The first block contains only 200 distinct values and is compressed. The second block in this example is not compressed since it contains too many distinct values to take advantage of the additional dictionary of indirect encoding.

7.4.1 Example

Given is the dictionary-encoded attribute vector for the first name column (5 million distinct values) of the world population table that is sorted by country. The number of bits required to store 5 million distinct values is 23 bit ($\lceil log_2(5$ million$)\rceil =$ 23 bit). Thus, the size of this vector without additional compression is 21.4 GB (8 billion $\cdot$ 23 bit).

Now we split up the attribute vector into blocks of 1,024 elements resulting in 7.8 million blocks ($\lceil \frac{8\ billion\ rows}{1024\ elements} \rceil$). For our calculation and for simplicity, we assume that each set of 1,024 people of the same country contains on average 200 different first names and all blocks will be compressed. The number of bits required to represent 200 different values is 8 bit ($\lceil log_2(200)\rceil = 8$ bit). As a result, the elements in the compressed attribute vector need only 8 bit instead of 23 bit when using local dictionaries.

Dictionary sizes can be calculated from the (average) number of distinct values in a block (200) multiplied by the size of the corresponding old valueID (23 bit) being the value in the local dictionary. For the reconstruction of a certain row, a pointer to the local dictionary for the corresponding block is stored (64 bit). Thus, the runtime for accessing a row is constant. The total amount of memory necessary for the compressed attribute vector is calculated as follows:

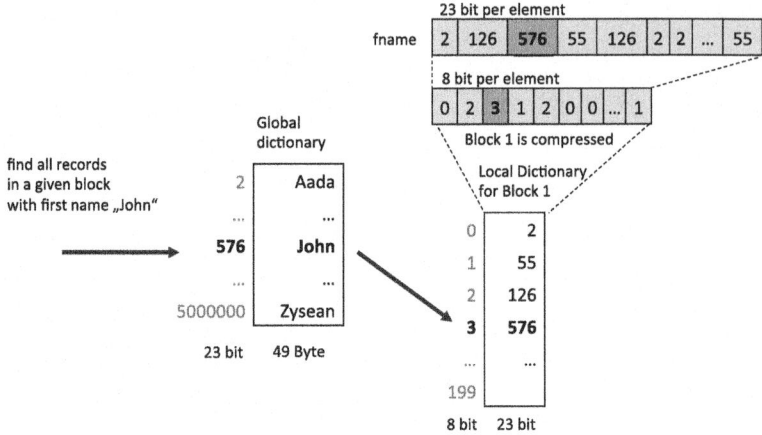

Fig. 7.6 Indirect encoding example

$$local\ dictionaries + compressed\ attribute\ vector$$

$$= (200 \cdot 23\ bit + 64\ bit) \cdot 7.8\ million\ blocks + 8\ billion \cdot 8\ bit$$

$$= 4.2\ GB + 7.6\ GB$$

$$\approx 11.8\ GB$$

Compared to the 21.4 GB for the dictionary-encoded attribute vector, a saving of 9.6 GB (44 %) can be achieved in our example.

The following example query that selects the birthdays of all people named "John" in the "USA" shows that indirect encoding allows for direct access:

```
SELECT birthday
FROM world_population
WHERE fname = 'John' AND country = 'USA'
```

Listing 7.1 Birthdays for all residents of the USA with first name John

As the table is sorted by country, we can easily identify the recordIDs of the records with country="USA", and determine the corresponding blocks to scan the "fname" column by dividing the first and last recordID by the cluster size. Then, the valueID for "John" is retrieved from the global dictionary and, for each block, the global valueID is translated into the local valueID by performing a binary search on the local dictionary. This is illustrated in Fig. 7.6 for a single block. Then, the block is scanned for the local valueID and corresponding recordIDs are returned for the birthday projection. In most cases, the starting and ending recordID will not match the beginning and the end of a block. In this case, we only consider the elements between the first above found recordID in the starting block up to the last found recordID for the value "USA" in the ending block.

7.5   Delta Encoding

The compression techniques covered so far reduce the size of the attribute vector. There are also some compression techniques to reduce the data volume in the dictionary as well. Let us assume that the data in the dictionary is sorted alphanumerically and we often encounter a large number of values with the same prefixes. Delta encoding exploits this fact and stores common prefixes only once.

Delta encoding uses a block-wise compression like in previous sections with typically 16 strings per block. At the beginning of each block, the length of the first string, followed by the string itself, is stored. For each following value, the number of characters used from the previous prefix, the number of characters added to this prefix and the characters added are stored. Thus, each following string can be composed of the characters shared with the previous string and its remaining part. Figure 7.7 shows an example of a compressed dictionary. The dictionary itself is shown in Fig. 7.7a. Its compressed counterpart is provided in Fig. 7.7b.

7.5.1   Example

Given is a dictionary for the city column sorted alpha-numerically. The size of the uncompressed dictionary with 1 million cities, each value using 49 Byte (we assume the longest city name has 49 letters), is 46.7 MB.

For compression purposes, the dictionary is separated into blocks of 16 values. Thus, the number of blocks is 62,500 ($\frac{1\ million\ cities}{16}$). Furthermore, we assume the following data characteristics to calculate the required size in memory:

- average length of city names is 7
- average overlap of 3 letters
- the longest city name is 49 letters ($\lceil log_2(49) \rceil = 6$ bit).

The size of the compressed dictionary is now calculated as follows:

$$block\ size \cdot number\ of\ blocks$$
$$= encoding\ lengths + 1^{st}\ city + 15\ other\ cities \cdot number\ of\ blocks$$
$$= ((1 + 15 \cdot 2) \cdot 6\ bit + 7 \cdot 1\ Byte + 15 \cdot (7 - 3) \cdot 1\ Byte) \cdot 62,500$$
$$\approx 5.4\ MB$$

Compared to the 46.7 MB without compression the saving is 42.2 MB (90 %).

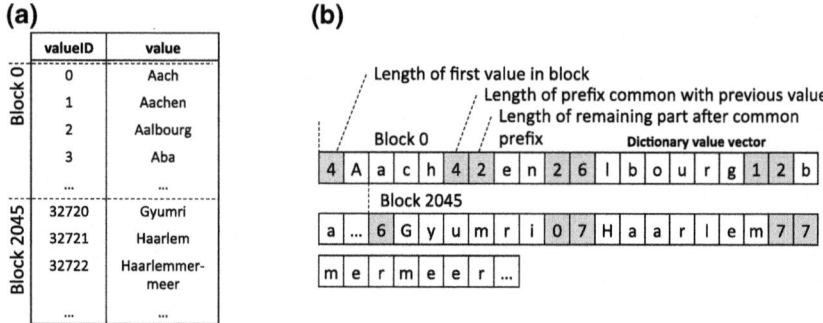

Fig. 7.7 Delta encoding example: (**a**) Dictionary, (**b**) Compressed dictionary

7.6   Limitations

What has to be kept in mind is that most compression techniques require sorted sets to tap their full potential, but a database table can only be sorted by one column or cascadingly if no other auxiliary data structures are used. Furthermore, some compression techniques do not allow direct access. This has to be carefully considered with regard to response time requirements of queries.

7.7   Self Test Questions

1. Sorting Compressed Tables
Which of the following statements is correct?

(a) If you sort a table by the amount of data for a row, you achieve faster read access
(b) Sorting has no effect on possible compression algorithms
(c) You can sort a table by multiple columns at the same time
(d) You can sort a table only by one column

2. Compression and OLAP / OLTP
What do you have to keep in mind if you want to bring OLAP and OLTP together?

(a) You should not use any compression techniques because they increase CPU load
(b) You should not use compression techniques with direct access, because they cause major security concerns
(c) Legal issues may prohibit to bring certain OLTP and OLAP datasets together, so all entries have to be reviewed
(d) You should use compression techniques that give you direct positional access, since indirect access is too slow

3. **Compression Techniques for Dictionaries**
   Which of the following compression techniques can be used to decrease the size of a sorted dictionary?

   (a) Cluster Encoding
   (b) Prefix Encoding
   (c) Run-Length Encoding
   (d) Delta Encoding

4. **Compression Example Prefix Encoding**
   Suppose there is a table where all 80 million inhabitants of Germany are assigned to their cities. Germany consists of about 12,200 cities, so the valueID is represented in the dictionary via 14 bit. The outcome of this is that the attribute vector for the cities has a size of 140 MB. We compress this attribute vector with Prefix Encoding and use Berlin, which has nearly 4 million inhabitants, as the prefix value. What is the size of the compressed attribute vector?
   Assume that the needed space to store the amount of prefix values and the prefix value itself is neglectable, because the prefix value only consumes 22 bit to represent the number of citizens in Berlin and additional 14 bit to store the key for Berlin once. Further assume the following conversions: 1 MB = 1,000 kB, 1 kB = 1,000 B

   (a) 0.1 MB
   (b) 133 MB
   (c) 63 MB
   (d) 90 MB

5. **Compression Example Run-Length Encoding Germany**
   Suppose there is a table where all 80 million inhabitants of Germany are assigned to their cities. The table is sorted by city. Germany consists of about 12,200 cities (represented by 14 bit). Using Run-Length Encoding with a start position vector, what is the size of the compressed city vector? Always use the minimal number of bits required for any of the values you have to choose and include all needed auxiliary structures. Further assume the following conversions: 1 MB = 1,000 kB, 1 kB = 1,000 B

   (a) 1.2 MB
   (b) 127 MB
   (c) 5.2 kB
   (d) 62.5 kB

6. **Compression Example Cluster Encoding**
   Assume the world population table with 8 billion entries. This table is sorted by countries. There are about 200 countries in the world. What is the size of the attribute vector for countries if you use Cluster Encoding with 1,024 elements per block assuming one block per country can not be compressed? Use the minimum required count of bits for the values and include all needed auxiliary structures.

Further assume the following conversions: 1 MB = 1,000 kB, 1 kB = 1,000 B

(a) ≈ 9 MB
(b) ≈ 4 MB
(c) ≈ 0.5 MB
(d) ≈ 110 MB

7. Best Compression Technique for Example Table
Find the best compression technique for the name column in the following table. The table lists the names of all inhabitants of Germany and their cities, i.e. there are two columns: first_name and city. Germany has about 80 million inhabitants and 12,200 cities. The table is sorted by the city column. Assume that any subset of 1,024 citizens contains at most 200 different first names.

(a) Run-Length Encoding
(b) Indirect Encoding
(c) Prefix Encoding
(d) Cluster Encoding

References

[AMF06]   D. Abadi, S. Madden, M. Ferreira, Integrating compression and execution in column-oriented database systems, in *Proceedings of the 2006 ACM SIGMOD International Conference on Management of Data , SIGMOD '06* (ACM, New York 2006), pp. 671–682
[LSFZ10] C. Lemke, K.-U. Sattler, F. Faerber, A. Zeier, Speeding up queries in column stores, in *Data Warehousing and Knowledge Discovery*. Lecture Notes in Computer Science, vol. 6263 (Springer, Heidelberg, 2010), pp. 117–129

Chapter 8
Data Layout in Main Memory

In this chapter, we address the question how data is organized in memory. Relational database tables have a two-dimensional structure but main memory is organized unidimensional, providing memory addresses that start at zero and increase serially to the highest available location. The database storage layer has to decide how to map the two-dimensional table structures to the linear memory address space.

We will consider two ways of representing a table in memory, called row and columnar layout, and a combination of both ways, a hybrid layout.

8.1 Cache Effects on Application Performance

In order to understand the implications introduced by row-based and column-based layouts, a basic understanding of memory access performance is essential. Due to the different available types of memory as described in Sect. 4.1, modern computer systems leverage a so-called memory hierarchy as described in Sect. 4.2. These caching mechanisms plus techniques like the Translation Lookaside Buffer (TLB, see Sect. 4.4) or hardware prefetching (see Sect. 4.5) introduce various performance implications, which will be outlined in this section that is based on [SKP12].

The described caching and virtual memory mechanisms are implemented as transparent systems from the viewpoint of an actual application. However, knowing the used system with its characteristics and optimizing applications based on this knowledge can have crucial implications on application performance.

The following two sections describe two small experiments, outlining performance differences when accessing main memory. These experiments are for the interested reader and will not be relevant for the exam.

H. Plattner, *A Course in In-Memory Data Management*,
DOI 10.1007/978-3-642-55270-0_8, © Springer-Verlag Berlin Heidelberg 2014

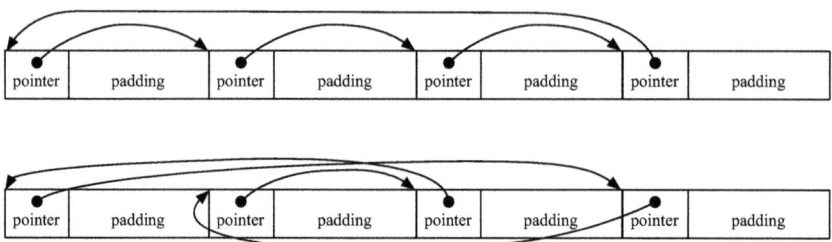

Fig. 8.1 Sequential vs. random array layout

8.1.1 The Stride Experiment

As the name random access memory suggests, the memory can be accessed randomly and one would expect constant access costs. In order to test this assumption, we run a simple benchmark accessing a constant number of addresses with an increasing stride, i.e. distance, between the accessed addresses.

We implemented this benchmark by iterating through an array chasing a pointer. The array is filled with structs. Structs are data structures, which allow to create user-defined aggregate data types that group multiple individual variables together. The structs consist of a pointer and an additional data attribute realizing the padding in memory, resulting in a memory access with the desired stride when following the pointer chained list.

```
struct element {
    struct element *pointer;
    size_t padding[PADDING];
}
```

Listing 8.1 Definition of the element structure for the stride experiment

In case of a sequential array, the pointer of element i points to element $i + 1$ and the pointer of the last element references the first element so that the loop through all array elements is closed. In case of a random array, the pointer of each element points to a random element of the array while ensuring that every element is referenced exactly once. Figure 8.1 outlines the created sequential and random arrays.

If the assumption holds and random memory access costs are constant, then the size of the padding in the array and the array layout (sequential or random) should make no difference when iterating over the array. Figure 8.2 shows the result for iterating through a list with 4,096 elements, while following the pointers inside the elements and increasing the padding between the elements. As we can clearly see, the access costs are not constant and increase with an increasing stride. We also see multiple points of discontinuity in the curves, e.g. the access times increase heavily up to a stride of 64 bytes and continue increasing with a smaller slope.

Figure 8.3 indicates that an increasing number of cache misses is causing the increase in access times. The first point of discontinuity in Fig. 8.2 is quite exactly the size of the cache lines of the test system. The strong increase is due to the fact, that with a stride smaller than 64 bytes, multiple list elements are located on

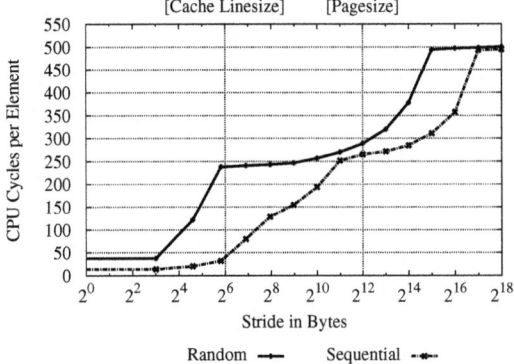

Fig. 8.2 Cycles for cache accesses with increasing stride

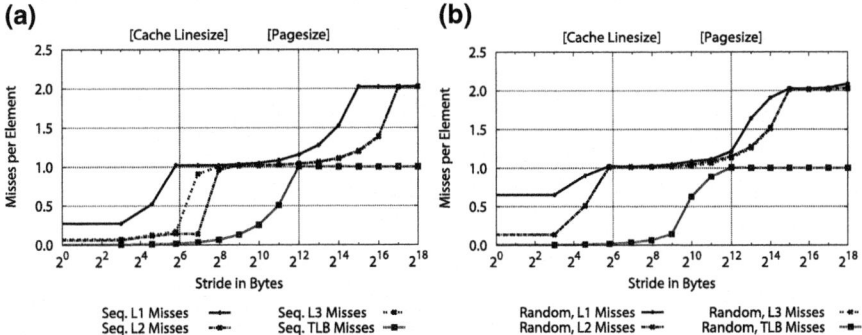

Fig. 8.3 Cache misses for cache accesses with increasing stride. (**a**) Sequential Access, (**b**) Random Access

one cache line and the overhead of loading one line is amortized over the multiple elements.

For strides greater than 64 bytes, we would expect a cache miss for every single list element and no further increase in access times. However, as the stride gets larger the array is placed over multiple pages in memory and more TLB misses occur, as the virtual addresses on the new pages have to be translated into physical addresses. The number of TLB cache misses increases up to the page size of 4 kB and stays at its worst case of one miss per element. With strides greater as the page size, the TLB misses can induce additional cache misses when translating the virtual to a physical address. These cache misses are due to accesses to the paging structures, which reside in main memory [BCR10, BT09, SS95].

To summarize, the performance of main memory accesses can largely differ depending on the access patterns. In order to improve application performance, main memory access should be optimized in order to exploit the usage of caches.

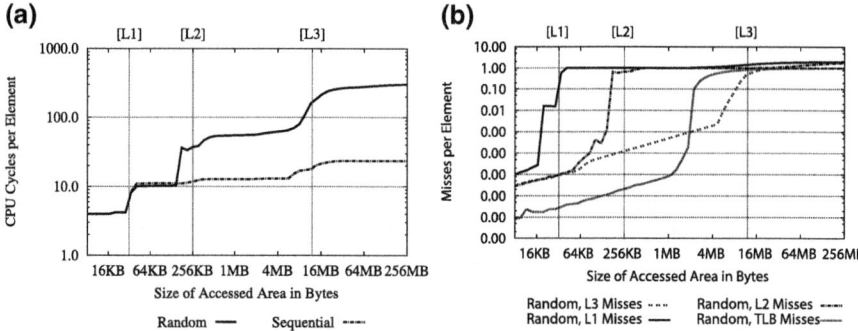

Fig. 8.4 Cycles and cache misses for cache accesses with increasing working sets. (**a**) Sequential Access, (**b**) Random Access

8.1.2 The Size Experiment

In a second experiment, we access a constant number of addresses in main memory with a constant stride of 64 bytes and vary the size of the working set size or accessed area in memory. A run with n memory accesses and a working set size of s bytes would iterate $\frac{n}{s \cdot 64}$ times through the array, which is created as described earlier in the stride experiment in Sect. 8.1.1.

Figure 8.4a shows that the access costs differ up to a factor of 100, depending on the working set size. The points of discontinuity correlate with the sizes of the caches in the system. As long as the working set size is smaller than the size of the L1 Cache, only the first iteration results in cache misses and all other accesses can be answered out of the cache. As the working set size increases, the accesses in one iteration start to evict the earlier accessed addresses, resulting in cache misses in the next iteration.

Figure 8.4b shows the individual cache misses with increasing working set sizes. Up to working sets of 32 kB, the misses for the L1 cache go up to one per element, the L2 cache misses reach their plateau at the L2 cache size of 256 kB and the L3 cache misses at 12 MB.

As we can see, the larger the accessed area in main memory, the more capacity cache misses occur, resulting in poorer application performance. Therefore, it is advisable to process data in cache-sized chunks if possible.

8.2 Row and Columnar Layouts

Let us consider a simple example to illustrate the two mentioned approaches for representing a relational table in memory. For simplicity, we assume that all values are stored as strings directly in memory and that we do not need to store any additional data. As an example, let us look at the simple world population example:

Id	Name	Country	City
1	Paul Smith	Australia	Sydney
2	Lena Jones	USA	Washington
3	Marc Winter	Germany	Berlin

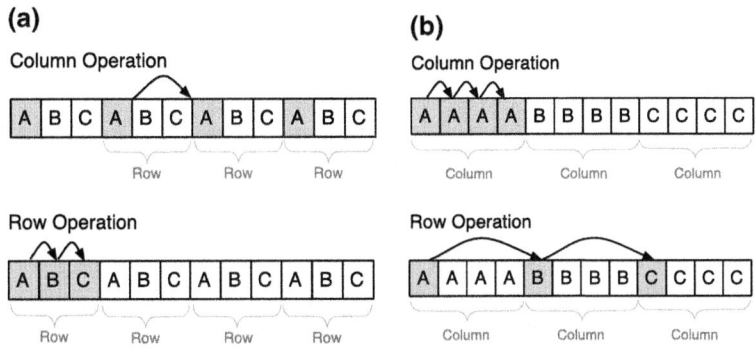

Fig. 8.5 Illustration of memory accesses for row-based and column-based operations on row and columnar data layouts. (**a**) Row Layout, (**b**) Columnar Layout

As discussed above, the database must transform its two-dimensional table into a one-dimensional series of bytes for the operating system to write them to memory. The classical and obvious approach is a row- or record-based layout. In this case, all attributes of a tuple are stored consecutively and sequentially in memory. In other words, the data is stored tuple-wise. Considering our example table, the data would be stored as follows: "1, Paul Smith, Australia, Sydney; 2, Lena Jones, USA, Washington; 3, Marc Winter, Germany, Berlin".

On the contrary, in a columnar layout , the values of one column are stored together, column by column. The resulting layout in memory for our example would be: "1, 2, 3; Paul Smith, Lena Jones, Marc Winter; Australia, USA, Germany; Sydney, Washington, Berlin".

The columnar layout is especially effective for set-based reads. In other words, it is useful for operations that work on many rows but only on a notably smaller subset of all columns, as the values of one column can be read sequentially, e.g. when performing aggregate calculations. However, when performing operations on single tuples or for inserting new rows, a row-based layout is beneficial. The different access patterns for row-based and column-based operations are illustrated in Fig. 8.5.

Currently, row-oriented architectures are widely used for OLTP workloads while column stores are widely utilized in OLAP scenarios like data warehousing, which typically involve a smaller number of highly complex queries over the complete data set.

8.3   Benefits of a Columnar Layout

As mentioned above, there are use cases where a row-based table layout can be more efficient. Nevertheless, many advantages speak in favor of the usage of a columnar layout in an enterprise scenario.

First, when analyzing the workloads enterprise databases are facing, it turns out that the actual workloads are more read-oriented and dominated by set processing (see Sect. 3.4 or [KKG+11]).

Second, despite the fact that hardware technology develops very rapidly and the size of available main memory constantly grows, the use of efficient compression techniques is still important in order to (a) keep as much data in main memory as possible and to (b) minimize the amount of data that has to be read from memory to process queries as well as the data transfer between non-volatile storage mediums and main memory.

Using column-based table layouts enables the use of efficient compression techniques leveraging the high data locality in columns (see Chap. 7). They mainly use the similarity of the data stored in a column. Dictionary encoding can be applied to row-based as well as column-based table layout, whereas other techniques like prefix encoding, run-length encoding, cluster encoding or indirect encoding only leverage their full benefits on columnar table layouts.

Third, using columnar table layouts enables very fast column scans as they can sequentially scan the memory, allowing e.g. on the fly calculations of aggregates. Consequently, storing pre-calculated aggregates in the database can be avoided, thus minimizing redundancy and complexity of the database.

8.4   Hybrid Table Layouts

As stated above, set processing operations are dominating enterprise workloads. Nevertheless, each concrete workload is different and might favor a row-based or a column-based layout. Hybrid table layouts combine the advantages of both worlds, allowing to store single attributes of a table column oriented while grouping other attributes into a row-based layout [GKP+11]. The actual optimal combination highly depends on the actual workload and can be calculated by layouting algorithms.

As an illustrating example, think about attributes, which inherently belong together in commercial applications, e.g. quantity and measuring unit or payment conditions in accounting. The idea of the hybrid layout is that if the set of attributes are processed together, it makes sense from a performance point of view to physically store them together. Considering the example table provided in Sect. 8.2 and assuming the fact that the attributes *Id* and *Name* are often processed together, we can outline the following hybrid data layout for the table: "1, Paul Smith; 2, Lena Jones; 3, Marc Winter; Australia, USA, Germany;

`Sydney, Washington, Berlin"`. In this case, *Id* and *Name* are stored row-based, but *country* and *city* are stored column-based. This hybrid layout may decrease the number of cache misses caused by the expected workload, resulting in increased performance.

The usage of hybrid layouts can be beneficial but also introduces new questions like how to find the optimal layout for a given workload or how to react on a changing workload.

8.5 Self Test Questions

1. **Consecutive Access vs. Stride Access**
   When DRAM can be accessed randomly with the same costs, why are consecutive accesses usually faster than stride accesses?

   (a) With consecutive memory locations, the probability that the next requested location has already been loaded in the cache is higher than with randomized/strided access. Furthermore is the memory page for consecutive accesses probably already in the TLB.
   (b) The bigger the size of the stride, the higher the probability, that two values are both in one cache line.
   (c) Loading consecutive locations is not faster, since the CPU performs better on prefetching random locations, than prefetching consecutive locations.
   (d) With modern CPU technologies like TLBs, caches and prefetching, all three access methods expose the same performance.

References

[BT09]   V. Babka, P. Tůma, Investigating cache parameters of x86 family processors, in *Proceedings of the 2009 SPEC Benchmark Workshop on Computer Performance Evaluation and Benchmarking* (Springer, Heidelberg, 2009), pp. 77–96

[BCR10] T.W. Barr, A.L. Cox, S. Rixner, Translation caching: skip, don't walk (the page table). ACM SIGARCH Comput. Archit. News **38**(3), 48–59(2010)

[GKP⁺11] M. Grund, J. Krueger, H. Plattner, A. Zeier, S. Madden, P. Cudre-Mauroux, HYRISE: a hybrid main memory storage engine, in *Proceedings of the VLDB Endowment*, vol. 4, issue 2 (November 2010), pp. 105–116

[KKG⁺11] J. Krueger, C. Kim, M. Grund, N. Satish, D. Schwalb, J. Chhugani, H. Plattner, P. Dubey, A. Zeier, Fast updates on read-optimized databases using multi-core CPUs, in *Proceedings of the VLDB Endowment*, vol. 5, issue 1 (September 2011)

[SS95]   R.H. Saavedra, A.J. Smith, Measuring cache and TLB performance and their effect on benchmark runtimes. IEEE Trans. Comput., (1995), pp. 1–79

[SKP12] D. Schwalb, J. Krueger, H. Plattner, Cache conscious column organization in in-memory column stores. Technical Report 67, Hasso Plattner Institute, December 2012

Chapter 9
Partitioning

9.1 Definition and Classification

Partitioning is the process of dividing a logical database into distinct independent datasets. Partitions are database objects itself and can be managed independently. The main reason to apply data partitioning is to achieve data-level parallelism and thus to enable performance gains. Nowadays, multi-core CPUs that are capable to process several distinct data areas in parallel harness partitioned data structures. Since partitioning is applied as a technical step to increase the query speed, it should be transparent[1] to the user. In order to ensure the transparency of partitioning for the end user, a view showing the complete table as a union of all query results from all involved partitions is required. Besides performance gains, data-level parallelism improves availability and manageability of datasets. Which of these sometimes contradicting goals is favored usually depends on the actual use case. Two examples are given in Sect. 9.4. Because data partitioning is a classical NP-complete[2] problem, finding the best partition is a complicated task, even if the desired goal has been clearly outlined [Kar72]. There are mainly two types of data partitioning: horizontal and vertical partitioning, which will be covered in detail in the following.

9.2 Vertical Partitioning

Vertical partitioning results in splitting the data into attribute groups with replicated primary keys. These groups are then distributed across two (or more) tables. Attributes that are usually accessed together should be in the same table, in order

---

[1]Transparent in IT means that something is completely invisible to the user, not that the user can inspect the implementation through the cover. Except of their effects like improvements in speed or usability, transparent components should not be noticeable at all.

[2]NP-complete means that the problem can not be solved in polynomial time.

H. Plattner, *A Course in In-Memory Data Management*,
DOI 10.1007/978-3-642-55270-0_9, © Springer-Verlag Berlin Heidelberg 2014

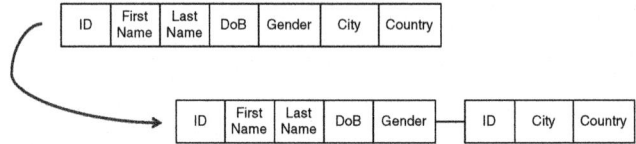

Fig. 9.1 Vertical partitioning

to increase join and materialization performance (Fig. 9.1). Such optimizations can only be applied if actual usage data exists, which is one point why application development should always be based on real customer data and workloads.

In row-based databases, vertical partitioning is possible in general, but is not a common approach. Column-based databases automatically support vertical partitioning, since each column can be regarded as a possible partition.

9.3   Horizontal Partitioning

Horizontal Partitioning is used more often in classic row-oriented databases. To apply this partitioning, the table is split into disjoint tuple groups by some condition. There are several sub-types of horizontal partitioning:

The first partitioning approach we present here is *range partitioning*, which separates tables into partitions by a predefined partitioning key, which determines how individual data rows are distributed to different partitions (Fig. 9.2). The partition key can consist of a single key column or multiple key columns. For example, customers could be partitioned based on their date of birth. If one is aiming for a number of four partitions, each partition would cover a range of about 25 years.[3] Because the implications of the chosen partition key depend on the workload, it is not trivial to find the optimal solution.

The second horizontal partitioning type is *round robin partitioning*. With round robin, a partitioning server does not use any tuple information as partitioning criteria, so there is no explicit partition key. The algorithm simply assigns tuples turn by turn to each partition, which automatically leads to an even distribution of entries and should support load-balancing to some extent (Fig. 9.3).

However, since specific entries might be accessed way more often than others, an even workload distribution can not be guaranteed. Improvements from intelligent data co-location or appropriate data-placement are not leveraged, because the data distribution is not dependent on the data, but only on the insertion order.

---

[3]Based on the assumption that the companies' customers mainly live nowadays and are between 0 and 100 years old.

Partition 1

ID	First Name	Last Name	DoB	Gender	City	Country

Partition 2

ID	First Name	Last Name	DoB	Gender	City	Country
3	Nina	Burg	1952/12/12	f	London	UK

Partition 3

ID	First Name	Last Name	DoB	Gender	City	Country
1	John	Dillan	1943/05/12	m	Berlin	Germany

Partition 4

ID	First Name	Last Name	DoB	Gender	City	Country
2	Peter	Black	1982/06/02	m	Austin	USA
4	Lucy	Sehan	1990/01/20	f	Jerusalem	Israel
5	Ariel	Shiva	1984/07/18	f	Tokio	Japan
6	Sharon	Lokida	1982/02/24	m	Madrid	Spain

Partitioning along the age: Partition 1:   76 – 100
Partition 2:   51 – 75
Partition 3:   26 – 50
Partition 4:   0 – 25

Fig. 9.2 Range partitioning

Partition 1

ID	First Name	Last Name	DoB	Gender	City	Country
1	John	Dillan	1943/05/12	m	Berlin	Germany
5	Ariel	Shiva	1984/07/18	f	Tokio	Japan

Partition 3

ID	First Name	Last Name	DoB	Gender	City	Country
3	Nina	Burg	1952/12/12	f	London	UK

Partition 2

ID	First Name	Last Name	DoB	Gender	City	Country
2	Peter	Black	1982/06/02	m	Austin	USA
6	Sharon	Lokida	1982/02/24	m	Madrid	Spain

Partition 4

ID	First Name	Last Name	DoB	Gender	City	Country
4	Lucy	Sehan	1990/01/20	f	Jerusalem	Israel

Fig. 9.3 Round robin partitioning

Partition 1

ID	First Name	Last Name	DoB	Gender	City	Country	hash(Country)
4	Lucy	Sehan	1990/01/20	f	Jerusalem	Israel	0x00

Partition 3

ID	First Name	Last Name	DoB	Gender	City	Country	hash(Country)
3	Nina	Burg	1952/12/12	f	London	UK	0x03

Partition 2

ID	First Name	Last Name	DoB	Gender	City	Country	hash(Country)
1	John	Dillan	1943/05/12	m	Berlin	Germany	0x01

Partition 4

ID	First Name	Last Name	DoB	Gender	City	Country	hash(Country)
2	Peter	Black	1982/06/02	m	Austin	USA	0x02
5	Ariel	Shiva	1984/07/18	f	Tokio	Japan	0x02

Fig. 9.4 Hash-based partitioning

The third horizontal partitioning type is *hash-based partitioning*. Hash partitioning uses a hash function[4] to specify the partition assignment for each row (Fig. 9.4).

The main challenge for hash-based partitioning is to choose a good hash function, that implicitly achieves locality or access improvements.

The last partitioning type is *semantic partitioning*. It uses knowledge about the application to split the data. For example, a database can be partitioned according

---

[4]A hash function maps a potentially large amount of data with often variable length to a smaller value of fixed length. In the figurative sense, hash functions generate a digital fingerprint of the input data.

to the life-cycle of a sales order. All tables required for the sales order represent one or more different life-cycle steps, such as creation, purchase, release, delivery, or dunning of a product. One possibility for suitable partitioning is to put all tables that belong to a certain life-cycle step into a separate partition.

9.4   Choosing a Suitable Partitioning Strategy

There are number of different optimization goals to be considered while choosing a suitable partitioning strategy. For instance, when optimizing for *performance*, it makes sense to have tuples of different tables, that are likely to be joined for further processing, on one server. This way the join can be done much faster due to optimal *data locality*, because there is no delay for transferring the data across the network. In contrast, for statistical queries like counts, tuples from one table should be distributed across as many nodes as possible in order to benefit from parallel processing.

To sum up, the best partitioning strategy depends very much on the specific use case.

9.5   Self Test Questions

1. **Partitioning Types**
   Which partitioning types do really exist and are mentioned in the course?

   (a) Selective Partitioning
   (b) Syntactic Partitioning
   (c) Range Partitioning
   (d) Block Partitioning

2. **Partitioning Type for Given Query**
   Which partitioning type fits best for the column 'birthday' in the world population table, when we assume that the main workload is caused by queries like 'SELECT first_name, last_name FROM population WHERE birthday >01.01.1990 AND birthday <31.12.2010 AND country = 'England'? Assume a non-parallel setting, so we can not scan partitions in parallel. The only parameter that is changed in the query is the country.

   (a) Round Robin Partitioning
   (b) All partitioning types will show the same performance
   (c) Range Partitioning
   (d) Hash Partitioning

3. Partitioning Strategy for Load Balancing
Which partitioning type is suited best to achieve fair load-balancing if the values of the column are non-uniformly distributed?

(a) Partitioning based on the number of attributes used modulo the number of systems
(b) Range Partitioning
(c) Round Robin Partitioning
(d) All partitioning types will show the same performance

Reference

[Kar72] R. Karp, Reducibility among combinatorial problems, in *Complexity of Computer Computations*, ed. by R. Miller, J. Thatcher (Plenum Press, New York, 1972), pp. 85–103

Part III
In-Memory Database Operators

Chapter 10
Delete

The delete operation terminates the validity of a given tuple. It stores the information in the database that a certain item is no longer valid. This operation can either be of *physical* or *logical* nature. A physical delete operation removes an item from the database so that it is no longer physically accessible. In contrast, a *logical* delete operation only terminates the validity of an item in the dataset, but keeps the tuple still available for temporal queries [Pla09].

The SQL syntax for a delete statement looks like the following, where the predicate may select a single or multiple tuples.

```
DELETE FROM table_name
WHERE condition
```

Listing 10.1 Delete syntax

10.1 Example of Physical Delete

In the following example, all persons with the name *'Jane Doe'* are supposed to be removed from a database table storing first and last names. Based on the applied dictionary encoding (see Chap. 6), the table consists of two columns with a dictionary and attribute vector each.

Dictionary "fname"		Attribute Vector "fname"		Dictionary "lname"		Attribute Vector "lname"	
valueID	value	recID	valueID	valueID	value	recID	valueID
...	...	...	...	...	...	...	...
22	Andrew	38	22	17	Brown	38	19
23	Jane	39	24	18	Doe	39	21
24	John	40	25	19	Miller	40	17
25	Mary	41	23	20	Schmidt	41	18
26	Peter	42	24	21	Smith	42	18
...	...	43	26	...	...	43	20
		...	...			...	...

H. Plattner, *A Course in In-Memory Data Management*,
DOI 10.1007/978-3-642-55270-0_10, © Springer-Verlag Berlin Heidelberg 2014

First, the valueIDs for the first and last name need to be identified. *Jane* corresponds to valueID 23 and *Doe* to valueID 18, according to their respective dictionary.

Dictionary "fname"		Attribute Vector "fname"		Dictionary "lname"		Attribute Vector "lname"	
valueID	value	recID	valueID	valueID	value	recID	valueID
...	...	...	...	...	...	...	...
22	Andrew	38	22	17	Brown	38	19
23	Jane	39	24	18	Doe	39	21
24	John	40	25	19	Miller	40	17
25	Mary	41	23	20	Schmidt	41	18
26	Peter	42	24	21	Smith	42	18
...	...	43	26	...	...	43	20
		...	...			...	...

Next, we scan through the attribute vectors and find the appropriate positions, which means we look up the recordIDs for these values. In our example, there is only one tuple with that combination of first and last name.

Dictionary "fname"		Attribute Vector "fname"		Dictionary "lname"		Attribute Vector "lname"	
valueID	value	recID	valueID	valueID	value	recID	valueID
...	...	...	...	...	...	...	...
22	Andrew	38	22	17	Brown	38	19
23	Jane	39	24	18	Doe	39	21
24	John	40	25	19	Miller	40	17
25	Mary	41	23	20	Schmidt	41	18
26	Peter	42	24	21	Smith	42	18
...	...	43	26	...	...	43	20
		...	...			...	...

When finally deleting the two values from the attribute vectors, all subsequent tuples need to be adjusted to maintain a sequence without gaps and they are moved to preserve a sequential memory area. This implementation alternative of the delete operation is therefore very expensive in terms of performance. In Chap. 26, later during the course, the *insert-only* approach is presented as a better alternative to implement deletion in typical enterprise use cases. This approach is of *logical* nature.

Dictionary "fname"		Attribute Vector "fname"		Dictionary "lname"		Attribute Vector "lname"	
valueID	value	recID	valueID	valueID	value	recID	valueID
...	...	...	...	...	...	...	...
22	Andrew	38	22	17	Brown	38	19
23	Jane	39	24	18	Doe	39	21
24	John	40	25	19	Miller	40	17
25	Mary	~~41~~	~~23~~	20	Schmidt	~~41~~	~~18~~
26	Peter	41	24	21	Smith	41	18
...	...	42	26	...	...	42	20
		...	...			...	...

10.2   Self Test Questions

1. **Delete Implementations**
   Which two possible delete implementations are mentioned in the course?

   (a) White box and black box delete
   (b) Physical and logical delete
   (c) Shifted and liquid delete
   (d) Column and row deletes

2. **Arrays to Scan for Specific Query with Dictionary Encoding**
   When applying a delete with two predicates, e.g. firstname = 'John' AND
   lastname = 'Smith', how many logical blocks in the IMDB are being looked
   at during determination which tuples to delete (all columns are dictionary
   encoded)?

   (a) 1
   (b) 2
   (c) 4
   (d) 8

3. **Fast Delete Execution**
   Assume a physical delete implementation and the following two SQL statements
   on our world population table:
   (A) DELETE FROM world_population WHERE country = 'China';
   (B) DELETE FROM world_population WHERE country = 'Ireland';
   Which query will execute faster? Please only consider the concepts learned so far.

   (a) Equal execution time
   (b) A
   (c) Depends on the ordering of the dictionary
   (d) B

Reference

[Pla09]   H. Plattner, A common database approach for OLTP and OLAP using an in-memory
          column database, in *SIGMOD*, pp. 1–2, 2009

Chapter 11
Insert

This chapter outlines what happens when inserting a new tuple into a table (execution of an insert statement). Compared to a row-based database, the insert in a column store is a bit more complicated. For a row-oriented database, the new tuple is simply appended to the end of the table, i.e., the tuple is stored as one piece. SanssouciDB uses column-orientation to store the data physically. A detailed description of the differences between row stores and column stores is given in Chap. 8. In a column store, adding a new tuple to the database means to add a new entry to every column that the table consists of. Internally, every column consists of a dictionary and an attribute vector (see Chap. 6). Adding a new entry to a column means to check the dictionary and adding a new value if necessary. Afterwards, the respective value of the dictionary entry is added to the attribute vector of the column. Since the dictionary is sorted, adding a new entry to a column results in three different scenarios:

1. Adding without a new dictionary entry
2. Adding with a new dictionary entry, without resorting the dictionary
3. Adding with a new dictionary entry, with resorting the dictionary

In this chapter, we will give a step by step explanation of the three different scenarios.

11.1 Example

In this example, we insert the data of a new person into the *world_population* table (see Fig. 11.1) that we used before. The example outlines what happens for the column *lname*, representing the last name of a person, and *fname*, representing the first name of a person.

H. Plattner, *A Course in In-Memory Data Management*,
DOI 10.1007/978-3-642-55270-0_11, © Springer-Verlag Berlin Heidelberg 2014

Example Table: world_population

recID	fname	lname	gender	country	city	birthday
0	Martin	Albrecht	m	GER	Berlin	08-05-1955
1	Michael	Berg	m	GER	Berlin	03-05-1970
2	Hanna	Schulze	f	GER	Hamburg	04-04-1968
3	Anton	Meyer	m	AUT	Innsbruck	10-20-1992
4	Sophie	Schulze	f	GER	Potsdam	09-03-1977
...	...	...	...	...	...	...

INSERT INTO world_population
VALUES (Karen, Schulze, f, GER, Rostock, 06-20-2012)

Fig. 11.1   Example database table named *world_population*

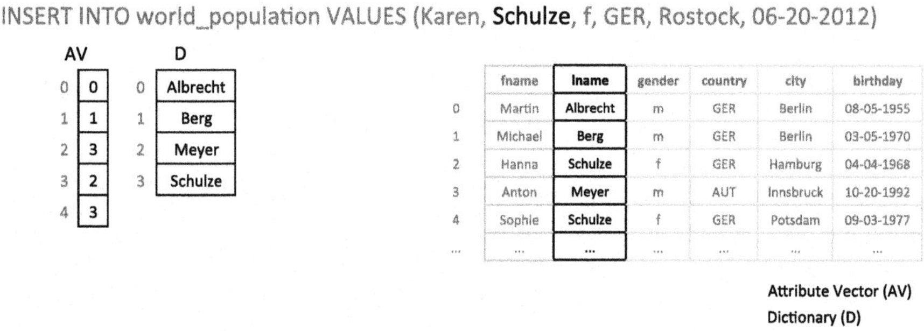

INSERT INTO world_population VALUES (Karen, **Schulze**, f, GER, Rostock, 06-20-2012)

Fig. 11.2   Initial status of the *lname* column

11.1.1   Inserting without New Dictionary Entry

To demonstrate a scenario were we have an insert without a new entry to the dictionary, we will look at the insert of the last name attribute to the *lname* column of our *world_population* table. Attribute vector and dictionary of the *lname* column are initially filled as displayed in Fig. 11.2.

To add the string *Schulze* to the column, we need to look up whether it already exists in the dictionary. Since there is another person named *Sophie Schulze* (recordID four of the world_population table) in the database, the dictionary for the *lname* column already contains an entry with the string *Schulze*. As one can see from Fig. 11.3, the dictionary position of *Schulze* is "3".

Since *Schulze* is on position 3 of the dictionary, we append 3 to the end of the attribute vector (see Fig. 11.4).

11.1.2   Inserting with New Dictionary Entry

When inserting the first name, the first name dictionary is scanned for the string *Karen*. As shown in Fig. 11.5, this name is not present in the dictionary, yet.

INSERT INTO world_population VALUES (Karen, **Schulze**, f, GER, Rostock, 06-20-2012)

AV		D	
0	0	0	Albrecht
1	1	1	Berg
2	3	2	Meyer
3	2	3	**Schulze**
4	3		

	fname	**lname**	gender	country	city	birthday
0	Martin	Albrecht	m	GER	Berlin	08-05-1955
1	Michael	Berg	m	GER	Berlin	03-05-1970
2	Hanna	Schulze	f	GER	Hamburg	04-04-1968
3	Anton	Meyer	m	AUT	Innsbruck	10-20-1992
4	Sophie	Schulze	f	GER	Potsdam	09-03-1977
...	...	...	...	...	...	...

Attribute Vector (AV)
Dictionary (D)

Fig. 11.3 Position of the string *Schulze* in the dictionary of the *lname* column

INSERT INTO world_population VALUES (Karen, **Schulze**, f, GER, Rostock, 06-20-2012)

AV		D	
0	0	0	Albrecht
1	1	1	Berg
2	3	2	Meyer
3	2	3	Schulze
4	3		
5	**3**		

	fname	**lname**	gender	country	city	birthday
0	Martin	Albrecht	m	GER	Berlin	08-05-1955
1	Michael	Berg	m	GER	Berlin	03-05-1970
2	Hanna	Schulze	f	GER	Hamburg	04-04-1968
3	Anton	Meyer	m	AUT	Innsbruck	10-20-1992
4	Sophie	Schulze	f	GER	Potsdam	09-03-1977
5		Schulze				
...	...	...	...	...	...	...

Attribute Vector (AV)
Dictionary (D)

Fig. 11.4 Appending valueID of *Schulze* to the end of the attribute vector

INSERT INTO world_population VALUES (**Karen**, Schulze, f, GER, Rostock, 06-20-2012)

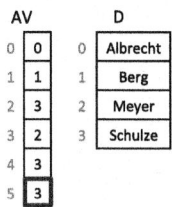

AV		D	
0	2	0	Anton
1	3	1	Hanna
2	1	2	Martin
3	0	3	Michael
4	4	4	Sophie

	fname	lname	gender	country	city	birthday
0	**Martin**	Albrecht	m	GER	Berlin	08-05-1955
1	**Michael**	Berg	m	GER	Berlin	03-05-1970
2	**Hanna**	Schulze	f	GER	Hamburg	04-04-1968
3	**Anton**	Meyer	m	AUT	Innsbruck	10-20-1992
4	**Sophie**	Schulze	f	GER	Potsdam	09-03-1977
5		Schulze				
...	**...**	...	...	...	...	...

Attribute Vector (AV)
Dictionary (D)

Fig. 11.5 Dictionary for first name column

Therefore, the name is appended to the end of the first name dictionary (see Fig. 11.6).

As outlined in Chap. 6, the dictionary needs to be kept sorted. After appending *Karen* to the end of the dictionary, the dictionary needs to be resorted. Therefore, as shown in Fig. 11.7, a new dictionary is created with a sorted order. In the new

INSERT INTO world_population VALUES (**Karen**, Schulze, f, GER, Rostock, 06-20-2012)

AV

0	2
1	3
2	1
3	0
4	4

D

0	Anton
1	Hanna
2	Martin
3	Michael
4	Sophie
5	Karen

	fname	lname	gender	country	city	birthday
0	Martin	Albrecht	m	GER	Berlin	08-05-1955
1	Michael	Berg	m	GER	Berlin	03-05-1970
2	Hanna	Schulze	f	GER	Hamburg	04-04-1968
3	Anton	Meyer	m	AUT	Innsbruck	10-20-1992
4	Sophie	Schulze	f	GER	Potsdam	09-03-1977
5		Schulze				
...	...	...	...	...	...	...

Attribute Vector (AV)
Dictionary (D)

Fig. 11.6 Addition of *Karen* to *fname* dictionary

INSERT INTO world_population VALUES (**Karen**, Schulze, f, GER, Rostock, 06-20-2012)

Attribute Vector (AV)
Dictionary (D)

Fig. 11.7 Resorting the *fname* dictionary

INSERT INTO world_population VALUES (**Karen**, Schulze, f, GER, Rostock, 06-20-2012)

Attribute Vector (AV)
Dictionary (D)

Fig. 11.8 Rebuilding the *fname* attribute vector

dictionary most of the values have been moved to a new position. For instance, the valueID for *Michael* changed from 3 to 4.

Based on the changed valueIDs of the new first name dictionary, all valueIDs of the first name attribute vector need to be updated as well. Figure 11.8 shows the changes to the attribute vector. For instance at position 1, the valueID for *Michael* is changed from 3 to 4.

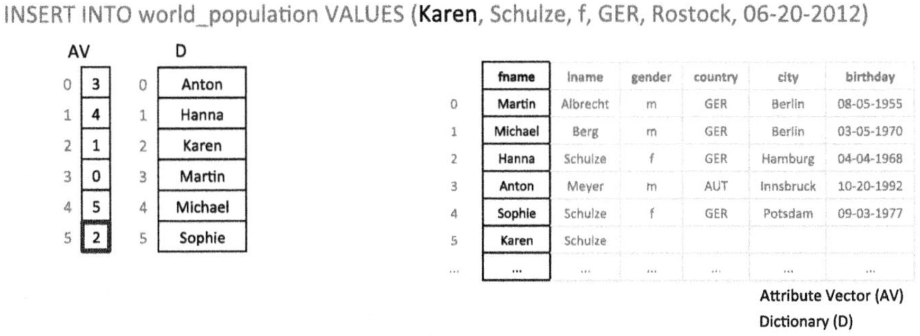

INSERT INTO world_population VALUES (**Karen**, Schulze, f, GER, Rostock, 06-20-2012)

	AV		D
0	3	0	Anton
1	4	1	Hanna
2	1	2	Karen
3	0	3	Martin
4	5	4	Michael
5	2	5	Sophie

	fname	lname	gender	country	city	birthday
0	Martin	Albrecht	m	GER	Berlin	08-05-1955
1	Michael	Berg	m	GER	Berlin	03-05-1970
2	Hanna	Schulze	f	GER	Hamburg	04-04-1968
3	Anton	Meyer	m	AUT	Innsbruck	10-20-1992
4	Sophie	Schulze	f	GER	Potsdam	09-03-1977
5	Karen	Schulze				
...	...	...	...	...	...	...

Attribute Vector (AV)
Dictionary (D)

Fig. 11.9 Appending the valueID representing *Karen* to the attribute vector

In case the newly added dictionary value is inserted at the end based on the sorting order of the dictionary, those two steps are omitted. The dictionary does not need to be resorted and therefore the attribute vector does not need to be rebuilt.

Finally the valueID 2, representing the dictionary position of the string *Karen*, is appended to the attribute vector (see Fig. 11.9).

11.2 Performance Considerations

When thinking of the *world_population* example, there are about 8 billion people and 5 million unique first names. Every new entry to the dictionary may cause an overhead regarding resorting of the dictionary and reorganization of the respective attribute vector. Triggering resorting and reorganization at every single insert would lead to a performance penalty, which compromises the overall performance of the system. Therefore, an additional insert layer needs to be added, the *differential buffer*. Chapter 25 explains in detail how write performance is kept at a high level using periodic merges of the differential buffer and the main store.

The vulnerability of a column to reorganization heavily depends on the column cardinality (the number of distinct values in a dictionary). When the dictionary only has a few entries, it is most likely that a column needs to be reorganized with a new insert. However, especially with attributes of low column cardinality, e.g., gender or country, the likelihood of reorganization decreases over time, since most of the possible values for the respective column have been inserted into the dictionary already. In real world applications, the dictionary only changes occasionally after it has reached a certain size. The additional steps necessary for new unique dictionary entries will occur less frequent and therefore expensive reorganization becomes less frequent.

11.3   Self Test Questions

1. **Access Order of Structures during Insert**
   When doing an insert, what entity is accessed first?

   (a) The attribute vector
   (b) The dictionary
   (c) No access of either entity is needed for an insert
   (d) Both are accessed in parallel in order to speed up the process

2. **New Value in Dictionary**
   Given the following entities:
   Old dictionary: ape, dog, elephant, giraffe
   Old attribute vector: 0, 3, 0, 1, 2, 3, 3
   Value to be inserted: lamb
   What value is the lamb mapped to in the new attribute vector?

   (a) 1
   (b) 2
   (c) 3
   (d) 4

3. **Insert Performance Variation over Time**
   Why might real world productive column stores experience faster insert performance over time?

   (a) Because the dictionary reaches a state of saturation and, thus, rewrites of the attribute vector become less likely.
   (b) Because the hardware will run faster after some run-in time.
   (c) Because the column is already loaded into main-memory and does not have to be loaded from disk.
   (d) An increase in insert performance should not be expected.

4. **Resorting Dictionaries of Columns**
   Consider a dictionary encoded column store (without a differential buffer) and the following SQL statements on an initially empty table: INSERT INTO students VALUES('Daniel', 'Bones', 'USA');
   INSERT INTO students VALUES('Brad', 'Davis', 'USA');
   INSERT INTO students VALUES('Hans', 'Pohlmann', 'GER');
   INSERT INTO students VALUES('Martin', 'Moore', 'USA');
   How often do attribute vectors have to be completely rewritten?

   (a) 2
   (b) 3
   (c) 4
   (d) 5

5. **Insert Performance**
   Which of the following use cases will have the worst insert performance when all values will be dictionary encoded?

   (a) A city resident database, that store all the names of all the people from that city
   (b) A database for vehicle maintenance data which stores failures, error codes and conducted repairs
   (c) A password database that stores the password hashes
   (d) An inventory database of a company storing the furniture for each room

Chapter 12
Update

The UPDATE operation is part of SQL's data manipulation language (DML) and is used to change existing tuples in a table. The UPDATE statement has the following syntax:

```
UPDATE table_name
SET column_name = value
[WHERE condition ]
```

Listing 12.1 Update Syntax

The optional WHERE condition restricts the update to tuples that match the given condition. If no WHERE condition is specified, then all tuples in the table are updated. Logically, i.e., in relational algebra, an UPDATE statement is equivalent to a DELETE statement followed by an INSERT statement.

12.1   Update Types

Three different types of updates can be found in a typical enterprise application [Pla09]:

- *Aggregate update:* The attributes are accumulated values as part of materialized views. From our experience in enterprise systems, typically between 1 and 5 materialized aggregates are maintained for each accounting line item.
- *Status update:* Binary change of a status variable, typically with timestamps
- *Value update:* The value of an attribute changes by replacement

H. Plattner, *A Course in In-Memory Data Management,*
DOI 10.1007/978-3-642-55270-0_12, © Springer-Verlag Berlin Heidelberg 2014

12.1.1   Aggregate Updates

Most of the updates taking place in financial applications are caused by updating aggregates. Such aggregates are used since computing them on the fly is too expensive in a row store without auxiliary data structures. An example for such an aggregate is the sum of sold items for each month and year. Now, whenever a new item is sold or a sold item is modified (e.g., the item is returned), the according aggregates have to be updated as well.

Materialized aggregates can be abandoned with the set processing performance of column stores such as SanssouciDB, because aggregates can be calculated on the fly. Without materialized aggregates, not only application logic becomes simpler but also data redundancy is reduced.

12.1.2   Status Updates

Status variables (e.g. unpaid, paid) typically use a predefined set of a small number of values and thus create no problem when performing an in-place update since the column cardinality does not change. It is advisable that compression of sequences (e.g. run-length encoding) in the columns is not allowed for status fields. If the automatic recording of status changes is preferable for the application, we can also use the insert-only approach, which will be discussed in Chap. 26, for these changes. In case the status variable has only two states, a null value and a time stamp can be used as values to note if the status has been set. Thus, an in-place update is fully transparent even considering temporal queries.

12.1.3   Value Updates

Since the change of an attribute in an enterprise application in most cases has to be recorded (log of changes), the insert-only approach (explained in detail in Chap. 26) seems to be the appropriate answer. On average only 5 % of the tuples of a financial accounting system are actually changed over a longer period of time [KKG+11]. The extra load for the differential buffer (the write-optimized store in SanssouciDB, which handles updates and inserts) and the extra consumption of main memory are acceptable. With insert-only, we also capture the change history including time and origin of the change.

Despite the fact that typical enterprise systems are not update-intensive, by using insert-only and by not maintaining totals, we can even further reduce the number of updates, which also reduces locking issues.

recID	fname	lname	gender	country	city	birthday
0	Martin	Albrecht	m	GER	Berlin	08-05-1955
1	Michael	Berg	m	GER	Berlin	03-05-1970
2	Hanna	Schulze	f	GER	Hamburg	04-04-1968
3	Anton	Meyer	m	AUT	Innsbruck	10-20-1992
4	Ulrike	Schulze	f	GER	Potsdam	09-03-1977
5	Sophie	Schulze	f	GER	Rostock	06-20-2012
...	...	...	...	...	...	...
8×10^9	Zacharias	Perdopolus	m	GRE	Athen	03-12-1979

Fig. 12.1 The *world_population* table before updating

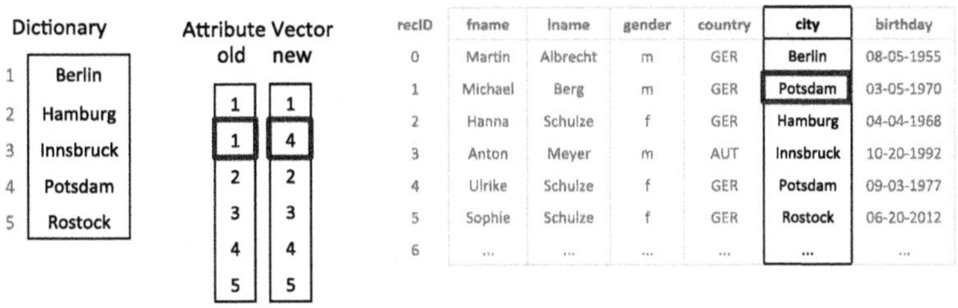

Fig. 12.2 Dictionary, old and new attribute vector of the *city* column, and state of the *world_population* table after updating

12.2 Update Example

Given is the world population table. Michael Berg moves from Berlin to Potsdam. So the following query should be executed:

```
UPDATE world_population SET city = 'Potsdam'
WHERE fname = 'Michael'
AND lname = 'Berg'
AND city = 'Berlin'
```

Listing 12.2 Michael Berg moves from Berlin to Potsdam

Figure 12.1 shows the table before the update is executed.

Because the value "Potsdam" already exists in the dictionary, the query executor can simply look up the dictionary position of "Potsdam" and update the attribute vector accordingly. This is shown in Fig. 12.2.

Now, assume that Hanna Schulze moves from Hamburg to Bamberg:

```
UPDATE world_population SET city = 'Bamberg'
WHERE fname = 'Hanna'
AND lname = 'Schulze'
AND city = 'Hamburg'
```

Listing 12.3 Hanna Schulze moves from Hamburg to Bamberg

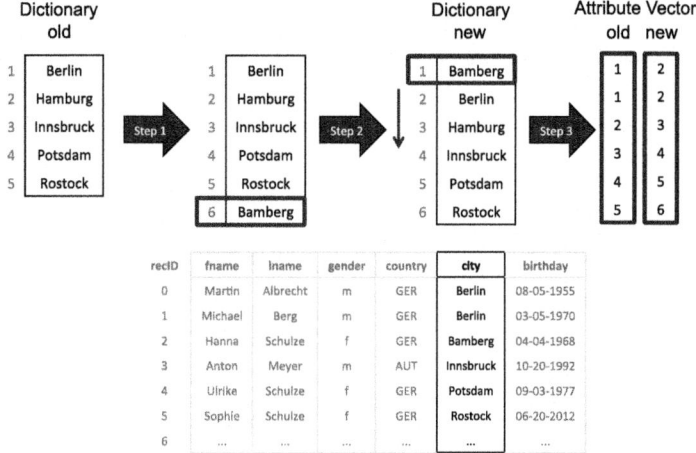

Fig. 12.3 Updating the *world_population* table with a value that is not yet in the dictionary

This time, the value "Bamberg" is not yet in the dictionary. Therefore, the query executor performs the following actions:

1. The value "Bamberg" is appended at the end of the dictionary.
2. The dictionary is reorganized in order to maintain its sort order, which is required for fast binary search on the dictionary.
3. Every value in the attribute vector is potentially updated (i.e. replaced with the new dictionary value representing the actual value). Depending on the position of the new value in the new-sorted dictionary, this step becomes very expensive. In our example, the complete attribute vector must be rewritten since "Bamberg" is the first item in the new-sorted dictionary. For such cases the performance considerations described in Sect. 11.2 are also true.

Figure 12.3 illustrates this process.

12.3 Self Test Questions

1. **Status Update Realization**
   How do we want to realize status updates for binary status variables?

   (a) Single status field: "false" means state 1, "true" means state 2
   (b) Two status fields: "true/false" means state 1, "false/true" means state 2
   (c) Single status field: "null" means state 1, a timestamp signifies transition to state 2
   (d) Single status field: timestamp 1 means state 1, timestamp 2 means state 2

2. **Value Updates**
   What is a "value update"?

   (a) Changing the value of an attribute
   (b) Changing the value of a materialized aggregate
   (c) The addition of a new column
   (d) Changing the value of a status variable

3. **Attribute Vector Rewriting After Updates**
   Consider the world population table (first name, last name) that includes all people in the world: Angela Mueller marries Friedrich Schulze and becomes Angela Schulze. Should the complete attribute vector for the last name column be rewritten?

   (a) No, because 'Schulze' is already in the dictionary and only the valueID in the respective row will be replaced
   (b) Yes, because 'Schulze' is moved to a different position in the dictionary
   (c) It depends on the position: All values after the updated row need to be rewritten
   (d) Yes, because after each update, all attribute vectors affected by the update are rewritten

References

[KKG⁺11]   J. Krueger, C. Kim, M. Grund, N. Satish, D. Schwalb, J. Chhugani, H. Plattner, P. Dubey, A. Zeier, Fast updates on read-optimized databases using multi-core CPUs. Proc. VLDB **5**, 61–72 (2011)

[Pla09]   H. Plattner, A common database approach for OLTP and OLAP using an in-memory column database, in *Proceedings of the SIGMOD International Conference on Management of Data*, ed. by U. Çetintemel, S.B. Zdonik, D. Kossmann, N. Tatbul (ACM, 2009), pp. 1–2

Chapter 13
Tuple Reconstruction

As mentioned earlier, data resembling a table can be stored in linear memory either column by column, i.e. columnar layout, or row by row, i.e. row layout. The impacts have already been discussed in Chap. 8 in more detail. The columnar layout is optimized for analytical set-based operations that work on many rows but for a notably smaller subset of all columns of data. The row layout shows a better performance for select operations on few single tuples. In this chapter, we discuss the operations needed for tuple reconstruction in detail and explain the influence of the different layouts on the performance of these operations. Tuple reconstruction is a typical functionality in OLTP applications. It is executed whenever more than one column is requested from the database, for example when the user in an ERP system calls the "show" or "edit" transactions for the master data object or for a document.

To explain the influence of the main memory layout organization on the performance of the tuple reconstruction operation, we have to consider the notion of the cache access and the size of the cache line. A CPU cache is a cache used by the central processing unit of a computer to reduce the average time to access memory. The cache is a smaller and faster memory which stores copies of the data from the most frequently used main memory locations. Memory cache is organized in 32 or 64 byte long cache lines. Even when reading just one byte from the memory, the CPU reads a complete cache line and places it into the cache. This characteristic of a cache will help us to estimate the response time for the tuple reconstruction operations for both layouts.

13.1 Tuple Reconstruction in Row-Oriented Databases

First, let us consider an example using the row layout. Let us assume we need to reconstruct the tuple when knowing the position of the tuple. As a first example, we take into account the following properties of the tuple:

H. Plattner, *A Course in In-Memory Data Management*,
DOI 10.1007/978-3-642-55270-0_13, © Springer-Verlag Berlin Heidelberg 2014

- the size of one tuple is 200 byte;
- the number of attributes in the tuple is 6.

To estimate the result, we also need the following parameters:

- speed of the read operation from main memory: 4 MB/ms/core;
- we consider 64 byte long cache lines;
- all calculations will be done for one core per CPU. If we consider more cores, the performance will increase appropriately.

Let us calculate how much time the read operation for the tuple reconstruction will take in this case considering that the data is organized using row layout. The operation is executed relatively fast, as all attributes are stored sequentially. Considering a size of 200 bytes per tuple, we will need 4 cache accesses ($\lceil \frac{200}{64} \rceil = 4$) to read the whole tuple from main memory. The CPU reads a bit more than the size of a tuple in this case, because it will read a complete cache line for every cache access. In case of a row layout, the CPU will load some data of the following tuple to the cache. Thus, we read 256 byte from main memory. Considering the speed 4 MB/ms/core, we can calculate the time as described below:

$$Tuple\ reconstruction\ response\ time\ (row\ layout) = \frac{256\ byte}{4,000,000\ byte/ms/core}$$

$$= 0.064\ \mu s$$

13.2   Tuple Reconstruction in Column-Oriented Databases

Now let us estimate the processing time for the same operation and tuples with the same characteristics but taking into account that the data is organized in a columnar layout. The data is stored attribute-wise in this case. To reconstruct the tuple, the CPU cannot just sequentially read data from memory. It needs to do cache accesses for every attribute required for tuple reconstruction. Therefore, knowing the implicit recordID of the tuple to be reconstructed, the CPU will "jump" between the attributes of the tuple to collect the values. Let us calculate how much time the read operation for the tuple reconstruction will take in this case. Considering that the reconstructed tuple has 6 attributes and that for a complete read of every attribute one cache access is required, we will need 6 cache accesses to read all attributes of the tuple from main memory. Taking into account a cache line size of 64 byte, the CPU needs to read: 64 byte · 6 = 384 byte from main memory. The CPU reads more than the tuple's size of 200 byte in this case, because it will read a complete cache line for every memory access. Using a columnar layout, the CPU will load some additional attributes' values of the following tuples. Considering the speed 4 MB/ms/core, we can calculate the time as described below:

$$\textit{Tuple reconstruction response time (column layout)} = \frac{384\ byte}{4,000,000\ byte/ms/core}$$

$$= 0.096\ \mu s$$

In this simple example, performance of the tuple reconstruction operation using a columnar layout is 50 % worse in comparison with the row layout. The difference in the response time can be even higher if we consider an example for a tuple with a larger number of attributes.

13.3 Further Examples and Discussion

In reality, the number of attributes in tables of business applications is much larger. As an example, let us calculate the response time for tuple reconstruction with the following characteristics:

- The size of one tuple is 3,200 byte. For the response time of the column layout calculation, we also consider that for every attribute of the tuple, one cache access is enough to read the whole attribute of the tuple.
- The number of attributes in the tuple is 100.

Let us calculate response times for the tuple reconstruction operation for both layouts considering the same CPU characteristics that were described in the example above.

Row layout: 50 cache accesses are required for a CPU to read the whole tuple: $50 \cdot 64$ byte $= 3,200$ byte.

$$\textit{Tuple reconstruction response time (row layout)} = \frac{3,200\ byte}{4,000,000\ byte/ms/core}$$

$$= 0.8\ \mu s$$

Columnar layout: 100 cache accesses are required in case of the columnar layout to read the attributes of the tuple: $100 \cdot 64$ byte $= 6,400$ byte.

$$\textit{Tuple reconstruction response time (column layout)} = \frac{6,400\ byte}{4,000,000\ byte/ms/core}$$

$$= 1.6\ \mu s$$

This example shows how the number of attributes of the tuple can influence the response time for both layouts. The performance for tuple reconstruction of the columnar layout will become progressively worse in comparison to the row store when we increase the number of a tuple's attributes and request all attributes.

We can conclude that it is important to select only the necessary fields of a tuple. This way, the potential disadvantage of tuple reconstruction using a columnar layout can be reduced to a minimum.

13.4   Self Test Questions

1. **Tuple Reconstruction on the Row Layout: Performance**
   Given a table with the following characteristics:

   - Physical storage in rows
   - The size of each field is 34 byte
   - The number of attributes is 9
   - A cache line has 64 byte
   - The CPU processes 4 MB per millisecond

   Calculate the time required for reconstructing a full row. Please assume the following conversions: $1\,MB = 1,000\,kB$, $1\,kB = 1,000\,B$

   (a) $\approx 0.05\,\mu s$
   (b) $\approx 0.125\,\mu s$
   (c) $\approx 0.08\,\mu s$
   (d) $\approx 0.416\,\mu s$

2. **Tuple Reconstruction on the Column Layout: Performance**
   Given a table with the following characteristics:

   - Physical storage in columns
   - The size of each field is 34 byte
   - The number of attributes is 9
   - A cache line has 64 byte
   - The CPU processes 4 MB per millisecond

   Calculate the time required for reconstructing a full row. Please assume the following conversions: $1\,MB = 1,000\,kB$, $1\,kB = 1,000\,B$

   (a) $\approx 0.08\,\mu s$
   (b) $\approx 0.725\,\mu s$
   (c) $\approx 0.144\,\mu s$
   (d) $\approx 0.225\,\mu s$

3. **Tuple Reconstruction in Hybrid Layout**
   A table containing product stock information has the following attributes:
   Warehouse (4 byte); Product Id (4 byte); Product Name Short (20 byte); Product Name Long (40 byte); Self Production (1 byte); Production Plant (4 byte); Product Group (4 byte); Sector (4 byte); Stock Volume (8 byte); Unit of Measure (3 byte); Price (8 byte); Currency (3 byte); Total Stock Value (8 byte); Stock Currency (3 byte)

The size of a full tuple is 114 byte.
The size of a cache-line is 64 byte.
The table is stored in main memory using a hybrid layout. The following fields are stored together:

- Stock Volume and Unit of Measure
- Price and Currency
- Total Stock Value and Stock Currency

All other fields are stored column-wise.
Calculate and select from the list below the time required for reconstructing a full tuple using a single CPU core with a scan speed of 4 MB per millisecond. Please assume the following conversions: 1 MB = 1,000 kB, 1 kB = 1,000 B

(a) ≈ 0.176 μs
(b) ≈ 0.04 μs
(c) ≈ 0.03 μs
(d) ≈ 0.213 μs

4. **Comparison of Performance of the Tuple Reconstruction on Different Layouts**

A table containing product stock information has the following attributes:
Warehouse (4 byte); Product Id (4 byte); Product Name Short (20 byte); Product Name Long (40 byte); Self Production (1 byte); Production Plant (4 byte); Product group (4 byte); Sector (4 byte); Stock Volume (8 byte); Unit of Measure (3 byte); Price (8 byte); Currency (3 byte); Total Stock Value (8 byte); Stock Currency (3 byte)
The size of a full tuple is 114 byte.
The size of a cache-line is 64 byte.
The scan speed of a CPU core is 4 MB per millisecond.
Which of the following statements are true?

(a) If the table is physically stored in column layout, the reconstruction of a single full tuple consumes ≈ 0.096 μs using a single CPU core.
(b) If the table is physically stored in row layout, the reconstruction of a single full tuple consumes ≈ 64 ns using a single CPU core.
(c) If the table is physically stored in column layout, the reconstruction of a single full tuple consumes ≈ 224 ns using a single CPU core.
(d) If the table is physically stored in row layout, the reconstruction of a single full tuple consumes ≈ 0.32 μs using a single CPU core.

Chapter 14
Scan Performance

14.1 Introduction

In this chapter, we discuss the performance of the scan operation. Scan operations search one or more attributes for a certain value. Unless the table is sorted by the attribute to scan, a scan has to iterate over all entries and returns those entries, which fulfill the search predicate (e.g. "SELECT * FROM world_population WHERE lastname = "Smith"). As in Chap. 13, we will discuss the influence of different layouts (row and column) and different approaches on the performance of the scan operation. We compare the following three approaches:

- full table scan in row layout
- stride access for the selected attributes in row layout
- full column scan in columnar layout

In the following calculations, the world population table already known from previous chapters is scanned. To recap, the table has the following properties:

- 8 billion tuples
- tuple size of 200 byte
- table size of 8 billions · 200 byte = 1.6 TB
- attributes: first name, last name, gender, country, city, birthday
- all attributes have a fixed length

In addition, the previous assumptions for the response time calculations are used:

- scan speed of read operations from main memory: 4 MB/ms/core
- cache line size of 64 byte

In the first example, the scan operation will help to answer the question: "How many women are in the world?".

The target column that has to be scanned to answer this question is "Gender", which has two possible distinct values. For simplicity, the calculations of the scan

H. Plattner, *A Course in In-Memory Data Management*,
DOI 10.1007/978-3-642-55270-0_14, © Springer-Verlag Berlin Heidelberg 2014

performance are done using a single core. When performing the scan operation, each row of the table is independent from all other rows. Consequently the scan operation can be efficiently parallelized and scales nearly linearly.

14.2   Row Layout: Full Table Scan

Having the data organized in a row layout, the first and most obvious approach to find the exact number of women in the world is to scan sequentially through all rows to read the gender attribute. We have seen this behavior for software that uses Object-Relational Mapping (ORM) and does calculations on the application side. Having to retrieve whole data sets in order to create the needed objects to interact with, results in a full table scan. During this operation, the CPU will read 1.6 TB from main memory. Taking into account the scan speed of 4 MB/ms per core, we can calculate the runtime on one core as follows:

$$Full\ table\ scan\ response\ time\ on\ 1\ core = \frac{1.6\ TB}{4\ MB/ms} = 400\ s$$

We would have to wait for more then 10 min to get the answer to our question. In order to achieve a better performance we have to look for optimizations. An obvious and simple solution is to compute the question in parallel on multiple cores and CPUs. We could do a vertical partitioning of the table and let the processing units execute the scan operation on the table parts in parallel.

Let us have a quick look on an example for a quad core CPU for that. The scan speed for a quad core CPU can be calculated as follows: 4 cores · 4 MB/ms/core = 16 MB/ms. The full table scan response time is 1.6 TB / 16 MB/ms = 100 s. Even with four cores the query execution takes more than a minute.

Another approach that could help to increase the performance of the scan operation is to take advantage of the in-memory database and read the gender fields with direct access. On disk based databases, only pages instead of single attributes are usually directly accessible (see Sect. 4.4). The results for this approach are calculated and discussed in the next Sect. 14.3.

14.3   Row Layout: Stride Access

Assume we still use a row layout to store the data in main memory. But now, instead of scanning the whole table and reading all table fields from main memory, we read the target field via direct access. This is possible when each row attribute has a fixed length, so that the distance between each scanned attribute can be calculated. To scan all gender fields the CPU does 8 billions cache accesses, one access for each tuple. Assuming the cache line size is 64 byte, and considering the fact that a CPU will read exactly 64 byte during each cache access independently of the gender field size, we can calculate the data size that is read from main memory during the whole scan operation:

$$data\ volume\ =\ 8\ billion \cdot 64\ byte \approx 512\ GB$$

Taking into account the scan speed, we get the following the response time for one core:

$$Stride\ access\ response\ time\ on\ 1\ core\ =\ 512\ GB\ /\ 4\ MB/ms\ =\ 128\ s$$

The result is better than that for the full table scan, but answering the question still takes several minutes. However, there are further opportunities to optimize the scan speed for our initial question. Section 14.4 will discuss the effects of using a columnar data layout.

14.4   Columnar Layout: Full Column Scan

When using a columnar layout, the data is stored attribute-wise in main memory. This fact leads us to the following conclusions:

- As values of the same attribute are stored consecutively, the probability that the next accessed item has already been loaded as part of the same cache line, is relatively high.
- As attributes of the same type are stored together, effective compression algorithms can be used to reduce the data volume that is stored in memory and that has to be transferred between main memory and CPU. The shorter the compressed values, the more values are stored within the same cache line.
- The sequential access patterns of the scan operation can be exploited by CPU prefetchers. While this holds true for both row and column stores, with a columnar data layout only items that are about to be scanned are loaded into the caches, therefore leading to better utilization.

Consequently, two aspects of columnar layouts can be leveraged: scanning only target fields and reading compressed values. Both aspects reduce the data volume transferred between main memory and CPU and consequently reduce response times.

Considering the gender column is dictionary-encoded as described in Chap. 6, only one bit is necessary to encode the two possible values 'm' and 'f'. As before, we can calculate the data volume to be read from main memory using the size of the attribute and the number of tuples: 8 billion · 1 bit ≈ 1 GB, which leads to a full column scan response time of 1 GB / 4 MB/ms/core = 0.25 s on one core.

The result shows a significant difference in the performance in comparison with both presented approaches for the row layout. Further taking into account the opportunity to use several cores and to execute the scan operation in parallel, using the column layout we can speed up the answer to our example-question even more. While our example query, which is using only one attribute and is posed against a vast number of tuples might not be the common use case, it can be stated that in analytical workloads the general circumstances are favorable for this approach, as we have already seen in Chap. 3. Queries against huge data volumes that operate on a small number of columns are characteristic for analytical enterprise applications.

14.5   Additional Examples and Discussion

In our previous example, we considered an almost "perfect" case. With 1-bit length, the gender attribute was compressed to the minimum for dictionary-encoded values. This fact decreased the data volume to be transferred between CPU and main memory. Of course, the outcome of the performance calculation depends on the size of the scanned fields. For larger fields, the CPU will need to scan through a higher data volume and less values will fit into one cache line.

To compare the results with another attribute, let us take the same table from the first example and calculate the response time for the full column scan operation on the column "Birthday". This column has considerably more distinct values than the "Gender" field.

Considering that every value (i.e. valueID of a compressed value in the "Birthday" column) has a size of 2 byte, we can calculate the transferred data volume and appropriate response time as follows:

- data volume to be read from main memory = 8 billion · 2 byte ≈ 16 GB
- full column scan response time = 16 GB / 4 MB/ms/core = 4 s (with one core)

To summarize the calculations performed above, we can conclude, that the following parameters of a CPU and a scanned table influence scan performance:

- cache utilization
- memory bandwidth
- number of processing units
- table cardinality (number of tuples in the table)
- used compression mechanism
- used layout: column or row layout

The example calculations in this chapter show a significant speed up of the scan performance when switching from row to dictionary-encoded column layout. While the columnar layout with its higher data density already better utilizes the CPU caches, we would also like to note that it enables further optimizations, e.g. the usage of SIMD/SSE operations (see Sect. 17.1.2).

14.6 Self Test Questions

1. **Loading Tuples from a Dictionary-Encoded Row-Oriented Layout**
   Consider the example presented in Sect. 14.2. We will now do the calculation assuming we have dictionary-encoded attributes. For this question, let us assume that each of the 8 billion encoded tuples has a total size of 32 byte. What is the time that a single core processor with a scan speed of 4 MB per millisecond needs to scan the whole world_population table if all data is stored in a dictionary-encoded row layout?

   (a) 128 s
   (b) 32 s
   (c) 48 s
   (d) 64 s

Chapter 15
Select

In this chapter, we describe how an application can extract data that was once stored in the database. The SELECT statement is a combination of multiple relational operations, mainly selection, projection, and Cartesian product. We focus on the implications of SanssouciDB's column-orientated data layout.

15.1 Relational Algebra

Three different basic operations of the relational algebra can be used to create the SELECT statement of SQL. These are the Cartesian product, the projection and the selection.

15.1.1 Cartesian product

The Cartesian product (or cross product) is a binary operation, taking two relations R_1 and R_2 to produce the result relation $R_1 \times R_2$. Those are having n_{R_1} and n_{R_2} attributes and a cardinality of $|R_1|$ and $|R_2|$. The $\times$ operator builds the set of all ordered pairs (r_1, r_2) with $r_1 \in R_1$, $r_2 \in R_2$. As a result, a new relation R_3 with $n_{R_3} = n_{R_1} + n_{R_2}$ and $|R_3| = |R_1| \cdot |R_2|$ tuples is returned. After both relations were combined, projections and selections can be applied to reduce the size of the result set. Database systems usually use join operations to reduce the size of intermediate results, as described in Chap. 19.

15.1.2 Projection

Projection is used to mask or permute the attributes of its input relation. Looking at the logical layout of a table, projection is a "vertical" operator. It can be written as $\pi_{j_1,\ldots,j_n}(R)$, with j_1 to j_n being an ordered sequence representing the attributes

H. Plattner, *A Course in In-Memory Data Management*,
DOI 10.1007/978-3-642-55270-0_15, © Springer-Verlag Berlin Heidelberg 2014

of R contained in the projections result. Using a column-oriented data layout, only columns that are part of the projection (and those attributes used in predicates, which are not necessarily projected) need to be read by the database. For that reason, query processing consumes fewer resources if only a subset of the entire set of attributes needs to be touched.

15.1.3   Selection and Selectivity

When the data stored within a relation needs to be filtered by some criteria, the selection operator is used. The selection, written as σ, is a "horizontal" operator. It evaluates an expression (predicate) p consisting of the parts a and b that are combined via a binary operation θ. While a and b can be attribute names, specified, or calculated values, θ represents any binary operation (e.g., equals, greater, smaller) that evaluates to "true" or "false". Only tuples of the relation with a positive evaluation ("true") of θ are included into the result set.

The term *selectivity* is used in two different contexts. If selectivity is used in the context of a specific predicate within a query it is called *predicate selectivity*. If the term appears with regards to an attribute or a column of a table, it is spoken of *attribute selectivity*.

Predicate selectivity describes the ratio of the amount of returned tuples in proportion to the total amount of scanned tuples. It is defined as:

$$selectivity_p = \frac{\sigma_p(R)}{|R|} \tag{15.1}$$

The result is a value between 1 (all tuples satisfy the predicate and are retrieved if the query is executed) and 0 (all tuples are dismissed). A "high value" (near 1) is regarded as "high selectivity", which means that only few values are dismissed, whereas a "low value" (near 0) stands for "low selectivity", which means that many values are dismissed. This can easily lead to confusion. Intuitively high selectivity would be understood as "highly selective", which means that only the "best" entries are kept and many tuples are dismissed. Low selectivity would be intuitively regarded as few dismisses. As the words high and low therefore often cause ambiguity in this context, the terms *strong* and *weak* are favorable. *Strong selectivity* denotes values near 0, whereas *weak selectivity* is used for values near 1.

The selectivity of an attribute reflects how selective the values within a given column are. It is the inverse of the column cardinality c_a. The attribute selectivity is defined as:

$$selectivity_a = \frac{1}{c_a} \tag{15.2}$$

The query optimizer uses the attribute selectivity as a hint to determine which attribute should be used first to reduce the amount of tuples.

id	fname	lname	country	gender
2394	Gianluigi	Buffon	Italy	m
3010	Lena	Gercke	Germany	f
3040	Mario	Balotelli	Italy	m
3949	Manuel	Neuer	Germany	m
4902	Lukas	Podolski	Germany	m
20102	Klaas-Jan	Huntelaar	Netherlands	m

Fig. 15.1 Example database table *world_population*

15.2 Data Retrieval

In most applications, SELECT is a commonly used command. The typical SQL SELECT statement can be defined as

$$\text{SELECT} \quad \pi_{j_1,...,j_n}(R)$$
$$\text{FROM} \quad R$$
$$\text{WHERE} \quad \sigma_{a\theta b}(R)$$

Because SQL presents a declarative description of the result requested from the database, an ordered set of execution steps is required to extract the data from the database, a so-called query execution plan. For each SQL query, multiple execution plans can exist that deliver the same results with differing performance. Query optimizers are used to calculate the cost of different query execution plans. Relying on cost models and heuristics used within the optimizer an effective plan is chosen. The goal is to reduce the size of the result set as early as possible, e.g., by

- applying selections as early as possible
- ordering sequential selections so that the most restrictive ones are executed first
- ordering joins corresponding to their tables' cardinalities (smallest tables are used first)

As a concrete example we use the table shown in Fig. 15.1 and execute the following SELECT statement that retrieves the first names and last names of male Italians from the *world_population* table:

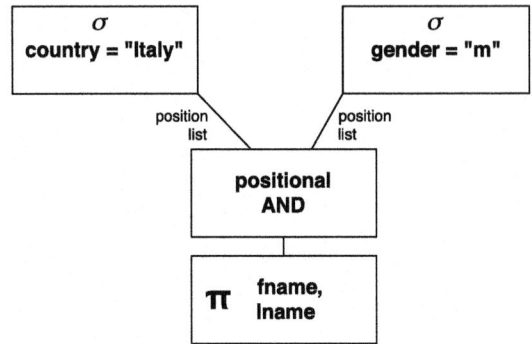

Fig. 15.2 Example query execution plan for SELECT statement

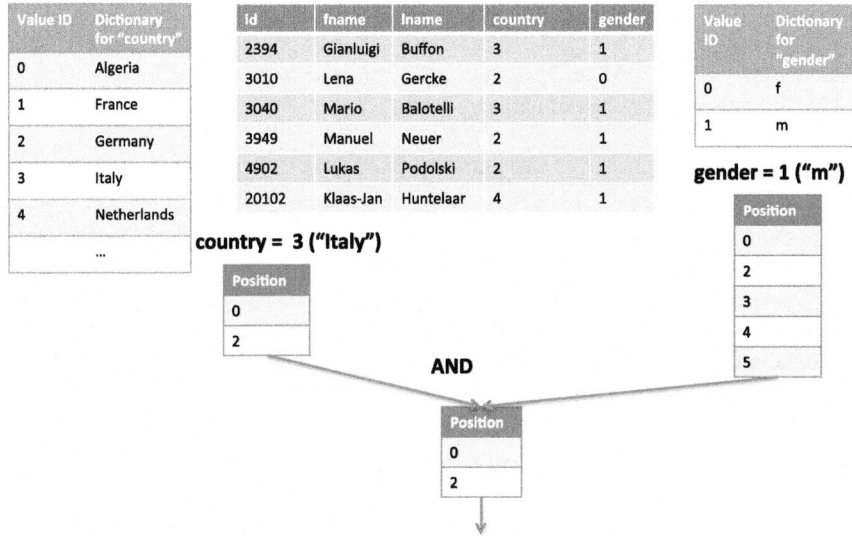

Fig. 15.3 Execution of the created query plan

```
SELECT fname , lname
FROM world_population
WHERE country = 'Italy' AND gender = 'm'
```

Listing 15.1 Example query

The corresponding query execution plan for that particular SQL query could look like shown in Fig. 15.2.

The query plan would than be executed in the database, as shown in Fig. 15.3. Database operations with independent inputs can be executed in parallel.

Because of SanssouciDB's dictionary encoding, a dictionary lookup is used to find the valueIDs for "Italy" and "m", in our example 3 and 1. Afterwards the attribute vectors of the country and gender columns are scanned and position lists identifying valid tuples are created. Those lists are intersected, resulting in a new list containing the positions of all tuples fulfilling the two selections. Finally, only the attributes *fname* and *lname* of the positionlist are projected, materialized and returned as the query result.

15.3 Self Test Questions

1. **Optimizing SELECT**
   How can the query optimizer improve the performance of SELECT statements?

   (a) By reducing the number of indices
   (b) By using the FAST SELECT keyword
   (c) Optimizers try to keep intermediate result sets large for maximum flexibility during query processing
   (d) By ordering multiple sequential select statements from strong (low) selectivity to weak (high) selectivity

2. **Selection Execution Order**
   Given is a query that selects the names of all German women born in the last 10 years from a world_population table, that stores information about all humans born in the last 100 years. In which order should the query optimizer execute the selections at best? Assume a sequential query execution plan and use the predicate selectivities for your calculations.

   (a) gender first, country second, birthday last
   (b) country first, gender second, birthday last
   (c) country first, birthday second, gender last
   (d) birthday first, gender second, country last

3. **Selectivity Calculation**
   Given is the query to select the names from German men born after January 1, 1990 and before December 31, 2009 from the world population table (8 billion people). Calculate the selectivity.
   Selectivity = number of tuples selected / number of tuples in the table
   Assumptions:

   • there are about 80 million Germans in the table
   • males and females are equally distributed in each country (50/50)
   • there is an equal distribution between all generations from 1910 until 2010

   (a) 0.001
   (b) 0.005
   (c) 0.1
   (d) 1

Chapter 16
Materialization Strategies

SQL is the most common language to interact with databases. Users are accustomed to the table-oriented output format of SQL. To provide the same data interfaces as known from row stores in column stores, the returned results have to be transformed into tuples in row format. The process of transforming encoded columnar data into tuples is called materialization.

Especially for column-oriented databases with lightweight compression, an appropriate materialization strategy is essential. Abadi et al. [AMDM07] analyzed different materialization strategies for column-oriented databases. Depending on the storage technique (e.g. compressed vs. uncompressed data, dictionary encoding vs. no dictionary encoding), different materialization strategies can be superior. Grund et al. [GKK+11] analyzed database operators and the impact of materialization strategies on intermediate results, in particular for dictionary-encoded columnar data structures.

16.1 Aspects of Materialization

Abadi et al. [AMDM07] divide the topic of materialization into two aspects, the execution of materialization and the time of materialization. The execution can be divided into parallel and pipelined materialization. The advantages and disadvantages of both approaches are discussed in detail by Grund et al. in [GKK+11]. All the following examples use a non-pipelined execution, where each operator is independent.

There are two different strategies concerning the time aspect of materialization: early and late materialization. Early materialization describes a strategy where data is decoded early (using dictionary lookups) during the query execution. For example, consider a dictionary-encoded string column. It consists of the attribute vector of integer values and the sorted dictionary of strings. Here, the actual string replaces the positional integer value representing the corresponding dictionary position early. Hence, a row-oriented tuple representation is created early on.

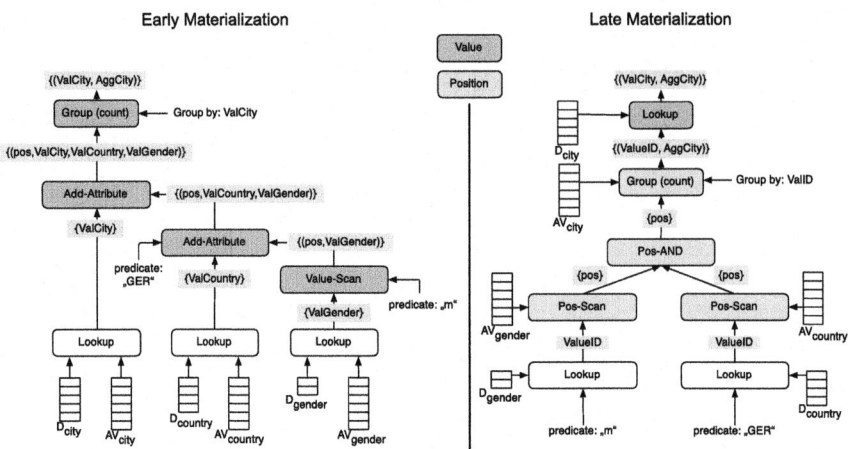

Fig. 16.1 Example comparison between early and late materialization

SELECT city, **COUNT**(*)
FROM world_population
WHERE gender = 'm'
  AND country = 'GER'
GROUP BY city

Listing 16.1 List the number of male inhabitants per city in Germany

With a late materialization strategy, column-orientation and the positional information instead of the actual value are used as long as possible during query execution. Ideally, the row-oriented tuple will be materialized in the very last step before returning the result to the user.

Figure 16.1 shows in an example where actual values and positions are used in early and late materialization.

In many cases, late materialization can improve the performance for column stores, especially when light-weight compression techniques are used [AMDM07]. The following sections will discuss both strategies based on an example query.

16.2 Example

To discuss the difference between early and late materialization, we will examine the query "List the number of male inhabitants per city in Germany", see the SQL query in Listing 16.1.

In both following examples, one strategy will be used throughout the whole query execution for exemplary purposes, even though a combination is often advantageous

Table "world_population"

fname	lname	gender	country	city	birthday
Martin	Albrecht	m	GER	Berlin	08-05-1955
Michael	Berg	m	GER	Berlin	03-05-1970
Hanna	Schulze	f	GER	Bonn	04-04-1968
Ulrich	Schulze	m	GER	Bonn	10-20-1992
...	...	...	...	...	...

Dictionary encoded attribute vectors

53946	10435	0	68	357	15556
54368	25063	0	68	357	20882
30145	99645	1	68	443	20182
99312	99645	0	68	443	29147
fname	lname	gender	country	city	birthday

Fig. 16.2 Example data of table *world_population*

in real world situations. Example data of the *world_population* table which is used in the query is shown in Fig. 16.2.

16.3 Early Materialization

When early materialization is used as the materialization strategy throughout the complete query, all required columns are materialized first. In our case, the required columns are all columns that are used as predicates in the query (*country* and *gender*), as well as all columns that are part of the result (i.e., *city*). Dictionary lookups are performed for each of these columns using the valueIDs in the corresponding attribute vectors. For the gender column, the result of these lookups is the vector {ValGender} with the actual values (see Fig. 16.3a).

The next step is to scan the intermediate vector {ValGender} for the gender predicate 'm'. To all qualifying lines the corresponding position is added and copied to the intermediate vector { (pos, ValGender) } (see Fig. 16.3b).

In the next step, the columns are combined as shown in Fig. 16.4. Hereby, the {ValCountry} vector is added to the intermediate result { (pos, ValGender) } while scanning for the predicate value 'GER'. Only tuples that are equal to the predicate are included in the result vector.

The final step is to aggregate and return the requested data of the SQL query. For that the intermediate result { (pos, ValGender, ValCountry, ValCity) } is grouped by ValCity and aggregated. The result is { (ValCity, AggCity) }, as shown in Fig. 16.5.

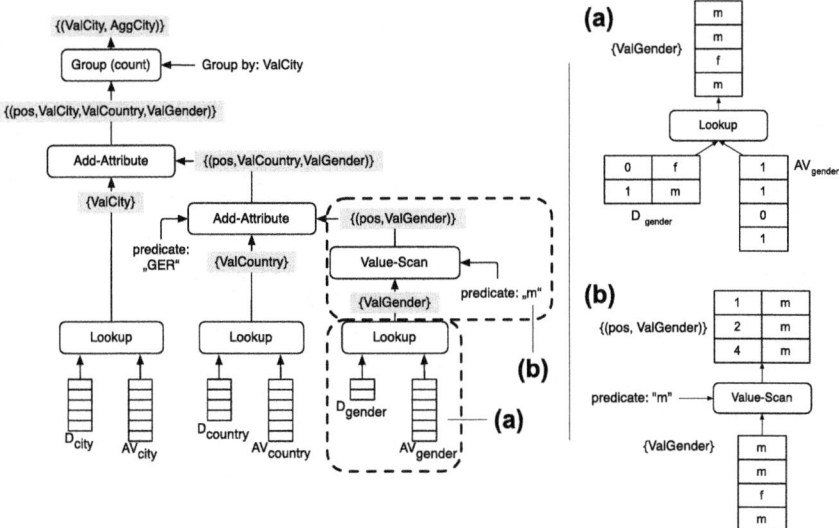

Fig. 16.3 Early materialization: materializing column via dictionary lookups and scanning for predicate

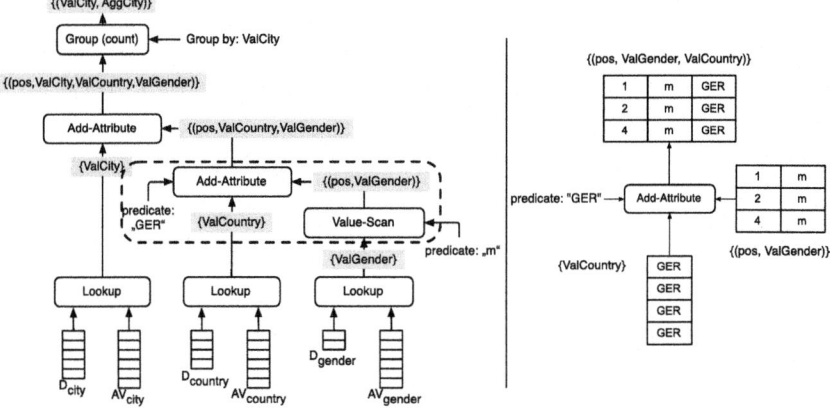

Fig. 16.4 Early materialization: scan for constraint and addition to intermediate result

16.4   Late Materialization

Instead of materializing the values of the dictionary lookup early (as done in the early materialization strategy), the dictionary-encoded value (valueID) contained in the attribute vector is being used. Ideally, the lookup into the dictionary for materialization is performed in the last step of the query execution before returning the result.

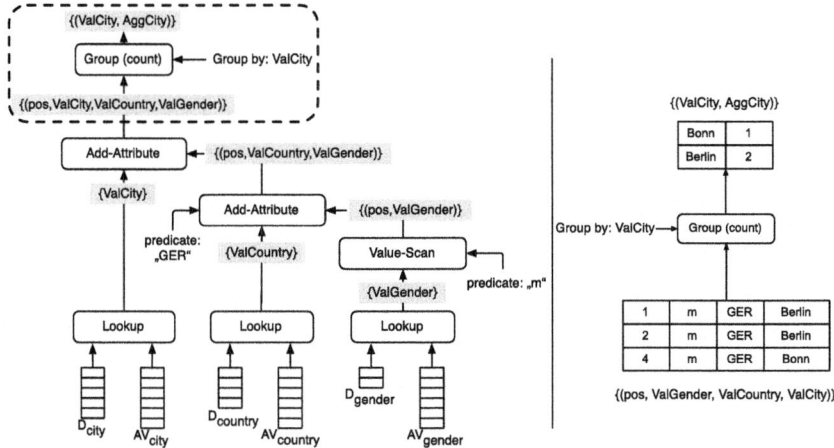

Fig. 16.5 Early materialization: group by *ValCity* and aggregation

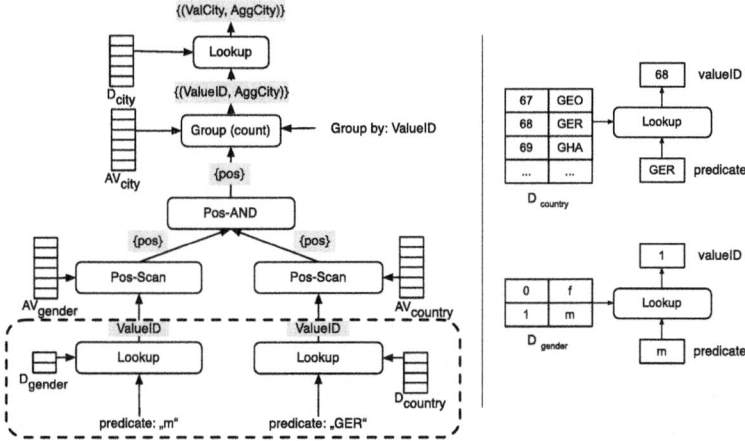

Fig. 16.6 Late materialization: lookup predicate values in dictionary

Figure 16.6 shows the first step. Here, the predicates *gender*='m' and *country*='GER' are used for the lookup using the corresponding dictionaries. The outcome is a vector of dictionary positions (valueIDs) per column that qualify for the given predicates. Notice that the dictionary for the column *city* is not accessed, since it is not required for the actual processing of the query right now. Only the valueID of the columns *gender* and *country* are looked up, as they are required for the succeeding scan operation.

Even though the visualization of the late materialization strategy implies a parallel execution of the lookups, the execution can also be done sequentially. Actually, with a predicate as *country*='GER', for which less than 2 % of the world population qualify, a sequential execution is advantageous (see Chap. 15).

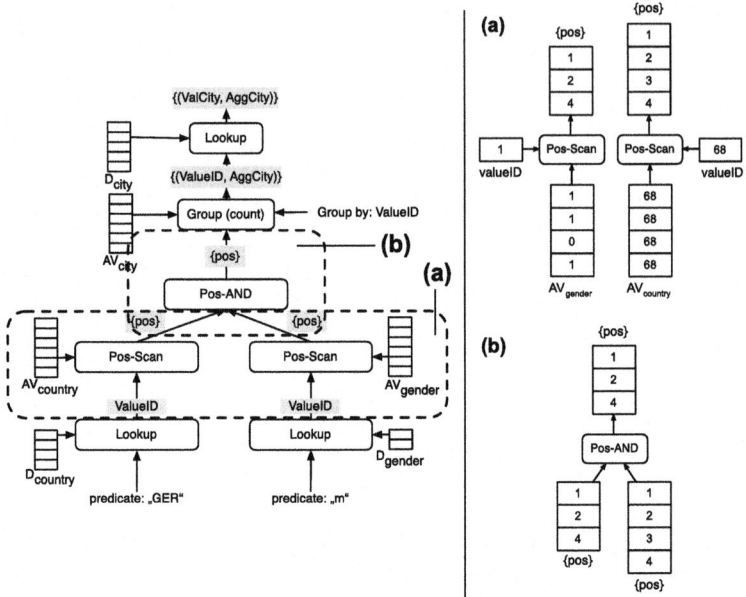

Fig. 16.7 Late materialization: scan and logical *AND*

Figure 16.7a shows the scan phase. With the valueIDs from the first step, now the attribute vectors are scanned. The position of each matching valueID in the attribute vector is added to the output vector of this step (*{pos}*). The merge of these positional lists is shown in Fig. 16.7b. Here, each value that occurs in both vectors is appended to the result vector of this step.

Figure 16.8a shows the group by operation. Hereby, the intermediate vectors are taken to group the positions in *{pos}* by the valueIDs in the city attribute vector and the count of each city is added to the output vector. In the last step the actual lookup of the city valueIDs is performed, as shown in Fig. 16.8b.

Compared to the early materialization strategy, the late materialization strategy might have to perform an additional lookup, e.g., when the gender would also be part of the result. This penalty can diminish the advantages, for example when many columns have to be materialized (consequently many dictionary lookups, what typically occurs when using 'SELECT *') or when the result set is very large (i.e., many output rows).

In general, the question to which extent—and if at all—late materialization is to be favored over early materialization depends on many variables like the used query operations and the selectivity [GKK+11].

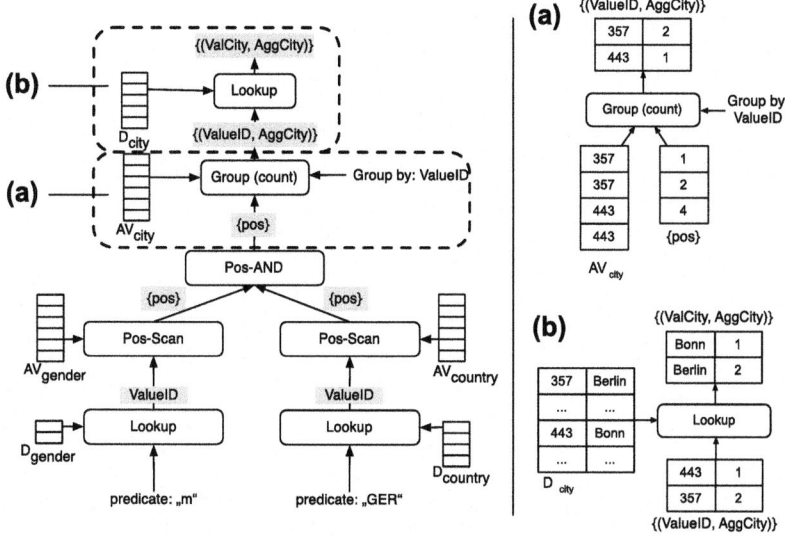

Fig. 16.8 Late materialization: filtering of attribute vector and dictionary lookup

16.5 Self Test Questions

1. **Performance of Materialization Strategies**
   Which materialization strategy—late or early materialization—provides the better performance?

   (a) Depends on the characteristics of the executed query
   (b) Late and early materialization always provide the same performance
   (c) Late materialization
   (d) Early materialization

2. **Characteristics of Early Materialization**
   Which of the following statements is true?

   (a) Whether late or early materialization is used is determined by the system clock
   (b) Early materialization requires lookups into the dictionary, which can be expensive and are not required when using late materialization
   (c) Depending on the persisted value types of a column, using positional information instead of actual values can be advantageous (e.g. in terms of cache usage or SIMD execution)
   (d) The execution of an early materialized query plan can not be parallelized

References

[AMDM07] D.J. Abadi, D. S. Myers, D.J. DeWitt, S. Madden, Materialization strategies in a
          column-oriented DBMS, in *ICDE*, Istanbul, ed. by R. Chirkova, A. Dogac, M.T.
          Özsu, T.K. Sellis (IEEE, 2007), pp. 465–475
[GKK⁺11] M. Grund, J. Krueger, M. Kleine, A. Zeier, H. Plattner, Optimal query operator
          materialization strategy for hybrid databases, in *IARIA*, pp. 169–174, 2011

Chapter 17
Parallel Data Processing

In the following, we discuss how to achieve parallelism in in-memory and traditional database management systems. Pipelined parallelism and data parallelism are two approaches to speed up query processing.

In pipelined parallelism, the next operator already starts while the current operator is not finished but has already produced partial results. Thus, the execution time of operators partly overlaps. For example, consider a JOIN and SORT operator involving the evaluation of a predicate. Each operator performs its tasks and if first results become available, they are used by the next operator as depicted in Fig. 17.1 on the left. This cascade of operations is denoted as a pipelining and can be extended without any limits.

In data parallelism, the data set is partitioned so that the operators of a query work on individual parts of the data set in parallel. Afterwards, the results of the parallel operations are merged to the complete result set (see Fig. 17.1 on the right). The query plan becomes more complex since all operators are executed on each data partition individually and a merge operation is added.

In database management systems, further aspects can be considered for parallelization. We distinguish between intra-query and inter-query parallelism. Intra-query parallelism addresses parallelization of operators within a query, i.e., the query looks like a single operation, but it is parallelized internally, e.g., by spawning multiple threads and using data parallelism. Inter-query parallelism addresses the aspect to schedule multiple queries to execute them in parallel. This results in efficient utilization of CPU caches and a reduction of data transfers.

17.1 Hardware Layer

Parallel processing of data is an essential aspect of achieving high performance for in-memory database systems. But what are the reasons for using parallelization instead of a single CPU core running at a tremendous high frequency, let us assume 1 PHz?

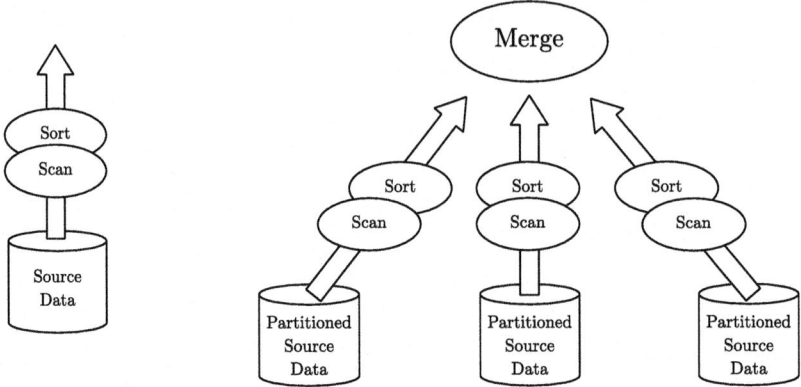

Pipelined Parallelism   **Data Parallelism**

Fig. 17.1 Pipelined and data parallelism

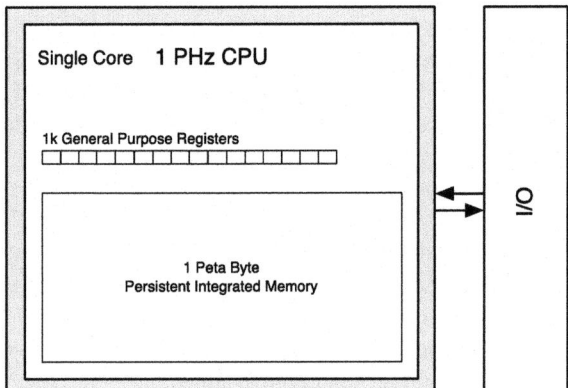

Fig. 17.2 The ideal hardware?

17.1.1 Multi-Core CPUs

Ideally, a modern computer would consist of a single CPU core running at 1 PHz and huge persistent integrated main memory as shown in Fig. 17.2. Reality looks different, though. Nowadays, we typically have multiple CPU cores on one CPU die. Furthermore, modern server systems consist of multiple CPUs. This multiplies the number of cores.

The reasons for that multi-core development are buried in the hardware developments of the last decade. The assumption that the number of transistors doubles every 18 month, known as Moore's law, is still valid [Moo65]. However, the operating frequency of the transistors cannot be increased infinitely. For example, with increasing frequency the ratio of heat loss increases. As a result, the energy

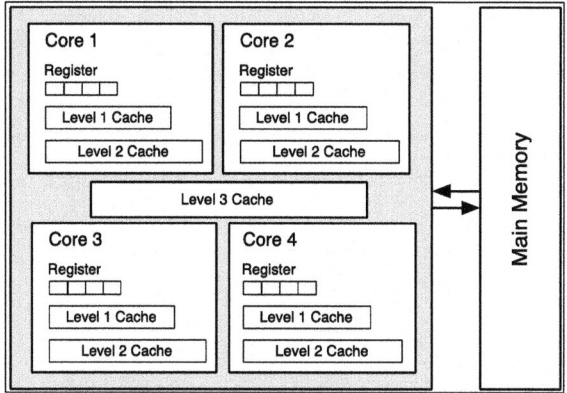

Fig. 17.3   A multi-core processor consisting of 4 cores

efficiency degrades while additional power is required to cool transistors. Hardware vendors proved that using multiple CPU cores operated at a lower frequency, e.g., 2.4–2.7 GHz, increases efficiency while keeping cooling requirements at an adequate level. For example, Fig. 17.3 depicts the conceptual architecture of single CPU consisting of four cores. Combining multiple CPUs within a single server is shown in Fig. 17.4 and the combination of multiple servers to form a more powerful data processing system is depicted in Fig. 17.5.

17.1.2   Single Instruction Multiple Data

The foundation of parallelization can directly be found within the CPU, i.e., data processing can be parallelized using the Single Instruction Multiple Data (SIMD) paradigm. SIMD allows to perform one data operation on multiple values in one CPU cycle. In contrast to traditional Reduced Instruction Set Computing (RISC) CPUs, SIMD parallelization builds on the use of so-called vectorized operators. These operators are directly implemented in the CPU to perform operations on multiple data words in specialized CPU registers in parallel. Computer graphics makes use of Streaming SIMD Extensions (SSE) instructions that operate on either 128 or 256 bit wide registers. For example, in one 128 bit register you can store four 32 bit values to perform a Parallel Add (PADD) as depicted in Fig. 17.6. Thus, four calculations can be processed within one instruction step instead of Scalar Add (SADD) where one calculation is performed at a time.

For instance, let us consider the aggregation of outstanding items. Using PADD reduces the time to sum up the individual items dramatically by summing up multiple items in a single instruction.

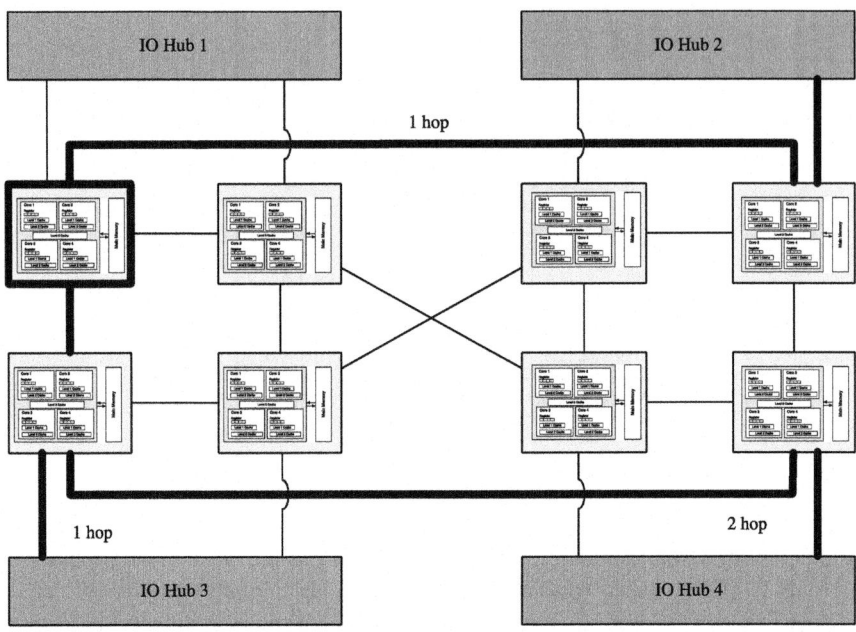

Fig. 17.4 A server consisting of multiple processors

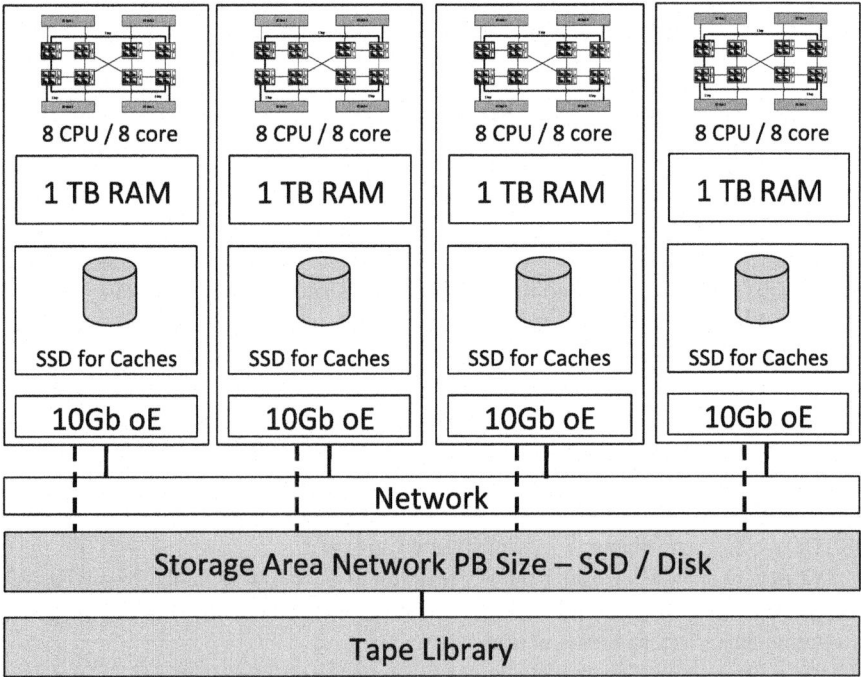

Fig. 17.5 A system consisting of multiple servers

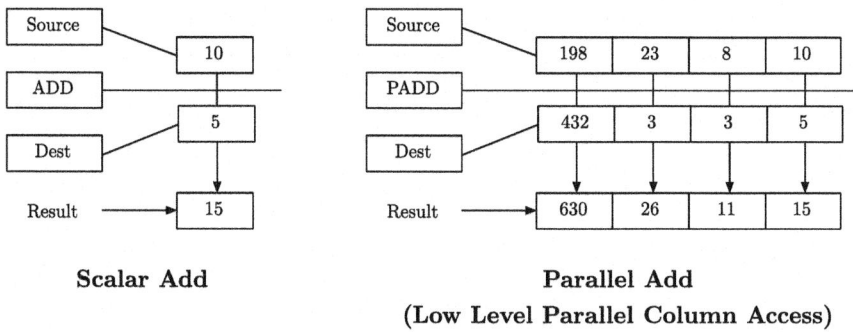

Scalar Add   **Parallel Add**
                        (Low Level Parallel Column Access)

Fig. 17.6 Single instruction multiple data parallelism

Let us assume the following example: the gender attribute values can be stored in a single bit. Using SIMD you can process the gender of 128 persons in a single CPU instruction step. For example, if you encode male as 1 and female as 0, calculating the ratio of male and female persons of this group of 128 persons is performed within a single instruction by performing a PADD. For comparison, modern processor families are able to perform 100,000 Million Instructions Per Second (MIPS) and more [HP11]. SIMD is the lowest level of parallelization on a computer system.

17.2 Software Layer

In addition to hardware parallelism, we consider software-level parallelism in the following section.

17.2.1 Amdahl's Law

Gene Amdahl conducted fundamental considerations about software-level parallelism. He defined that the maximum speedup of executing a piece of code in parallel is limited by the time needed to process the longest sequential fraction of the code. This principle is known as Amdahl's law [Amd67].

$$\text{max. speedup(N)} = \frac{1}{(1 - P) + \frac{P}{N}} \tag{17.1}$$

Equation (17.1) defines Amdahl's law with P giving the fraction of the code that can be processed in parallel and N giving the level of parallelism, e.g. the number of CPU cores.

Let us assume the following example: the ratio of parallel and the sequential part are 3:1. If the execution time of the parallel part is decreased, e.g. by increasing the number of cores, the maximum speedup cannot exceed four as can be seen in Eq. (17.2).

$$\lim_{N \to \infty} \text{speedup(N)} = \frac{1}{\left(1 - \frac{3}{4}\right) + \frac{3}{4N}} = \frac{1}{\frac{1}{4}} = 4 \qquad (17.2)$$

Amdahl's law assumes that the there is a fixed size of the solution space, i.e. a tasks generates repeatable a finite number of results. In contrast, Gustafson assumes that there is a maximum acceptable response time while the solution space is not known beforehand [Gus88]. Equation (17.3) defines Gustafson's law with C defining the number of cores and α defining the non-parallelizable fraction of the program code.

$$\text{max. speedup(C)} = C - \alpha \, (C - 1) \qquad (17.3)$$

17.2.2   Shared Memory

In a shared memory system [Li86], data that is stored in the shared memory segment is accessible by all processors in an uniform way. Special programming concepts, such as mutexes and semaphores, are used to avoid conflicting data access in the shared memory segment, e.g., simultaneous write access. Although shared memory is an easy way to share data across processes or CPU cores, it comes with the problem of scaling.

Shared memory systems suffer from scalability issue since the maximum size of the shared memory segment is limited by available memory size. The total memory size of a single system is small compared to the total main memory size formed by multiple servers.

17.2.3   Message Passing

Message passing is a very powerful paradigm to improve the processing of algorithmic problems [GLS94]. Instead of sharing memory between all threads, only messages are passed between individual processing threads. This paradigm is widely used for number crunching tasks, such as prediction of meteorology, earthquakes, and other kinds of simulations.

All CPU cores perform tasks independently while cores depending on results of each other exchange messages for coordination. In comparison to the shared memory approach, message passing can easily scale out, because processors are independent of shared memory. Thus, they can perform their tasks individually while exchanging messages, e.g., via network links. However, if the sum of exchanged messages exceeds the network capacity, the network becomes a bottleneck for this parallelism paradigm.

17.2.4   MapReduce

This data parallel paradigm aims at identifying a portion of data so that each portion can be processed in parallel. Thus, each processing job is performing the same task on an individual partition of the complete data. Examples of data parallel paradigm are the MapReduce framework and the OpenMP library [DG08, DM98].

MapReduce consists of two specific functions: the map and the reduce function. The former operates on individual data partitions in parallel and produces partial results $r_1 \ldots r_n$ for its assigned partition $p_1 \ldots p_n$. The reduce function forms one overall result r_{all} by merging all partial result $r_1 \ldots r_n$. Map and reduce steps can be chained to produce arbitrary results for complex tasks.

The canonical example for MapReduce is counting the number of occurrences of a specific word in a defined set of text documents. Each map function processes an individual text document or a part of it. It counts the number of occurrences for a specific word within this document. Since map functions are executed in parallel, multiple text documents are scanned for the desired word simultaneously. The following reduce function calculates the total number of occurrences for the specific word by summing up the individual results.

MapReduce requires the developer to define the "how" and the "what", i.e., if your algorithm does not scale efficiently, the overall response time will not be reduced. This direct control may also be a disadvantage for some tasks since you only want to define the "what". For example, in a database management system you expect an optimizer to generate the proper code—the "how"—to retrieve the desired data—the "what".

Thus, MapReduce does not address all problems efficiently. It is designed for parallel processing of a batch job, e.g., word counting. However, interactive analytical queries require flexible access to data. For example, exploring overdue payers requires subsequent analyses of data subsets, which makes it hard to have all possible map functions available.

17.3   Self Test Questions

1. **Amdahl's Law**
   Amdahl's Law states that . . .

   (a) the number of CPUs doubles every 18 months
   (b) the amount of available memory doubles every year

(c) the speedup of parallelization is limited by the time needed for the sequential fractions of the program

(d) the level of parallelization can be no higher than the number of available CPUs

2. Shared Memory
What limits the use of shared memory?

(a) The operation frequency of the processor
(b) The usage of Streaming SIMD instructions (SSE).
(c) The caches of each CPU
(d) The number of worker threads, which share the same resources and the limited memory itself.

References

[Amd67] G.M. Amdahl, Validity of the single processor approach to achieving large scale computing capabilities, in *Proceedings of the April 18–20, 1967, Spring Joint Computer Conference, AFIPS '67 (Spring)* (ACM, New York, 1967), pp. 483–485

[DM98]   L. Dagum, R. Menon, Openmp: an industry-standard api for shared-memory programming. IEEE Comput. Sci. Eng. **5**(1), 46–55 (1998)

[DG08]   J. Dean, S. Ghemawat, Mapreduce: simplified data processing on large clusters. Comm. ACM **51**(1), 107–113 (2008)

[GLS94] W. Gropp, E. Lusk, A. Skjellum, *Using MPI: Portable Parallel Programming with the Message-Passing Interface* (MIT Press, Cambridge, MA, 1994)

[Gus88] J.L. Gustafson, Reevaluating amdahl's law. Commun. ACM **31**(5), 532–533 (1988)

[HP11]   J.L. Hennessy, D.A. Patterson, *Computer Architecture: A Quantitative Approach*, 5th edn. (Elsevier Science, Burlington, 2011)

[Li86]   K. Li, Shared virtual memory on loosely coupled multiprocessors. Ph.D. thesis, New Haven, 1986 (AAI8728365)

[Moo65] G. Moore, Cramming more components onto integrated circuits. Electronics **38**, 114 ff. (1965)

Chapter 18
Indices

In row-oriented database systems indices are used to speed up access on columns that are often used for data selection (e.g. primary key attributes). Without indices a full table scan has to be performed when the data is stored row-wise. In column-oriented databases systems a data select operation on any attribute only requires a column scan. This is performed much faster than a complete table scan. Consequently, in many use cases additional indices are not required. The following chapter describes the usage of additional index structures in column-oriented tables, called inverted indices.

18.1 Indices: A Query Optimization Approach

Usually, applications work only with a subset of records at a time. Therefore, before processing the portion, it must be located within the database. Hence, records should be stored in a manner that makes it possible to locate them efficiently whenever they are needed. The process of locating a specific set of records is determined by the predicates that are used to characterize these records.

SanssouciDB organizes its records in columns (see Chap. 8). To determine the records, it is necessary to perform a scan on all the columns, which are used as filter criteria. In main memory column-oriented databases, which store column values continuously, i.e. in adjacent memory blocks, searching for a value with a full scan (by iterating through all items placed in memory sequentially) can be done orders of magnitude faster than in row-oriented databases. Therefore, the creation of additional index structures in such databases is not essential, such as in row-oriented systems. Nevertheless, because the complexity of a full column scan is linear ($\mathcal{O}(n)$), with the number of tuples, it is just a matter of data volume that will make the speed advantage of indices relevant to main memory column-oriented databases.

In this chapter, we discuss the topic of inverted indices in the context of main memory databases in more detail.

Example Table: world_population

recID	fname	lname	gender	country	city	birthday
0	Daniel	Specht	m	GER	Iserlohn	08-05-1955
1	Michael	Schardt	m	GER	Dresden	03-05-1970
2	Miriam	Faust	f	GER	Hamburg	04-04-1968
3	Toni	Meier	m	GER	Dresden	10-20-1992
4	Marlit	Mattes	f	GER	Berlin	09-03-1977
5	Tanja	Berger	f	GER	Aachen	07-10-1983
6	Thomas	Peters	m	GER	Berlin	04-07-1989
7	Ralf	Wagner	m	GER	Potsdam	01-11-1981

Fig. 18.1 Example database table named *world_population*

Column[city]

Attribute Vector (AV)   Dictionary (D)

	AV				D
0	4	Iserlohn	0		Aachen
1	2	Dresden	1		Berlin
2	3	Hamburg	2		Dresden
3	2	Dresden	3		Hamburg
4	1	Berlin	4		Iserlohn
5	0	Aachen	5		Potsdam
6	1	Berlin			
7	5	Potsdam			

Fig. 18.2 City column of the *world_population* table

```
SELECT * FROM world_population
WHERE city = 'Berlin';
```

Listing 18.1 Query to select all people from Berlin

18.2   Technical Considerations

In the *world_population* table shown in Fig. 18.1 we want to locate the records of all people from Berlin. The dictionary and the attribute vector of the column are shown in Fig. 18.2.

To determine the set of *Berlin* records, we need to set the filter criterion on the *city* attribute of the table. The respective SQL query could look like the one depicted in Listing 18.1.

Now, let us do a back of the envelope calculation. Assuming the table contains 8 billion records, our CPU is able to process 4 GB per second per core, and the city column is encoded using 20 bit. The memory footprint of the *city* column's attribute vector can be calculated using

$$8 \text{ billion} \cdot 20 \text{ bit} = 160 \text{ billion bit} = 20 \text{ billion Byte} \approx 18.6 \text{ GB}$$

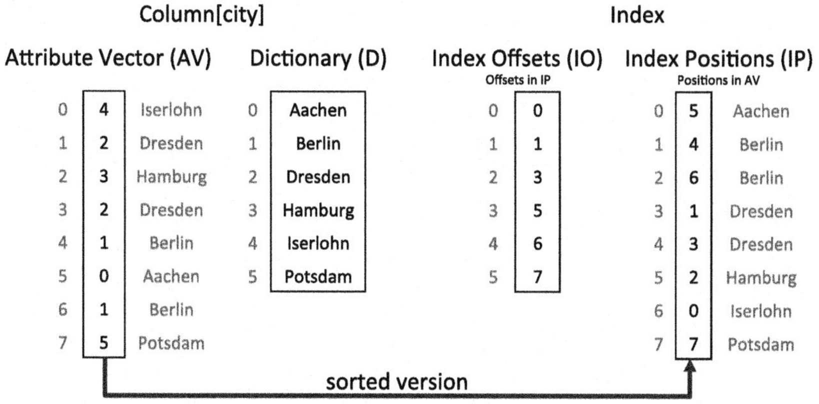

Column[city]   Index

Attribute Vector (AV)   Dictionary (D)   Index Offsets (IO)   Index Positions (IP)
                                                    Offsets in IP   Positions in AV

0	4	Iserlohn	0	Aachen	0	0	0	5	Aachen
1	2	Dresden	1	Berlin	1	1	1	4	Berlin
2	3	Hamburg	2	Dresden	2	3	2	6	Berlin
3	2	Dresden	3	Hamburg	3	5	3	1	Dresden
4	1	Berlin	4	Iserlohn	4	6	4	3	Dresden
5	0	Aachen	5	Potsdam	5	7	5	2	Hamburg
6	1	Berlin					6	0	Iserlohn
7	5	Potsdam					7	7	Potsdam

sorted version

Fig. 18.3 Index offset and index positions

The time it takes a single core to scan that amount of data can be calculated using

$$18.6\ GB \div 4\ GB/sec = 4{,}65\ sec$$

If the scan operation is parallelized between 40 cores, it takes $\approx 116\,\text{ms}$. While this speed is unthinkable for a row-oriented database, the speed might not be sufficient for all applications.

18.3   Inverted Index

Now, let us investigate a more complicated, but more efficient algorithm presented in detail in [FSKP12]. We consider an inverted index for the attribute *city*. An inverted index maps each distinct value to a position list, which contains all positions where the distinct value can be found in the column (Fig. 18.3). The index for the dictionary-encoded column consists of the following two parts:

- Index offsets (IO) : this vector stores for each dictionary entry (or in other words, for each unique value of the attribute vector) the offset for the position list in the positions vector. This means that the offset vector stores references to the first occurrence of the particular dictionary value in the positions vector.
- Index positions (IP) : the index position vector contains a position list of all distinct values of the attribute vector sorted by the integer valueID. In contrast, the attribute vector stores valueIDs by position.

Let us see how much data the CPU has to read using an index. We continue with the query shown in Listing 18.1. The following steps need to be executed to determine the position of *Berlin* in the attribute vector.

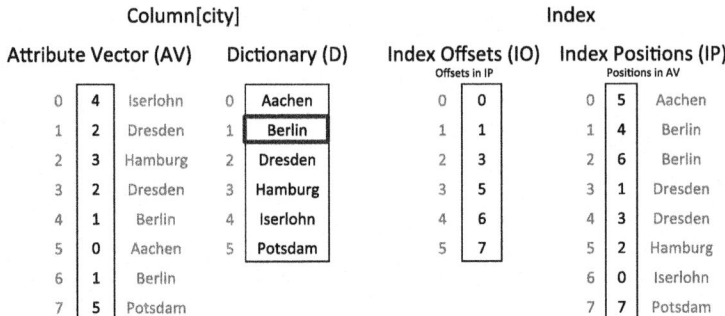

Fig. 18.4 Query processing using indices: Step 1

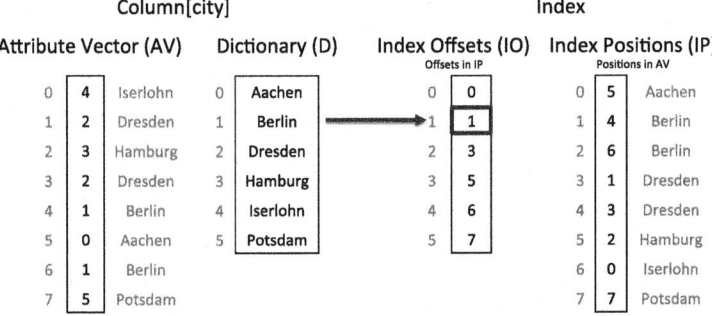

Fig. 18.5 Query processing using indices: Step 2

1. We need to perform a binary search on the dictionary to determine the dictionary position related to *Berlin*. As depicted in Fig. 18.4, *Berlin* is at position 1.
2. The dictionary position of *Berlin* corresponds directly to the position of the index offset vector shown in Fig. 18.5. In this example, the dictionary position of *Berlin* is 1, so the corresponding index offset vector position is 1.
3. Since the attribute of the respective search criterion is not necessarily a primary key, it is possible that the same value is used by many records. Consequently, more than one attribute vector entry can be filled with that value. As explained in the beginning of this chapter, the index position vector represents a sorted list of the values in the attribute vector. To determine the range of values to read from the index position vector, we simply read the value of the index offset vector at the position, we determined in Step 1, and the value of the next higher position (see Fig. 18.6).
4. Since 3 is already the offset of the next value in the index positions vector, we only need to read positions 1 and 2. As shown in Fig. 18.7, IP vector position 1 contains the value 4 and position 2 contains the value 6, which are the exact positions of the dictionary code for *Berlin* in the attribute vector. By retrieving the offsets of *Berlin* and the next entry, in our case *Dresden*, from the IO vector,

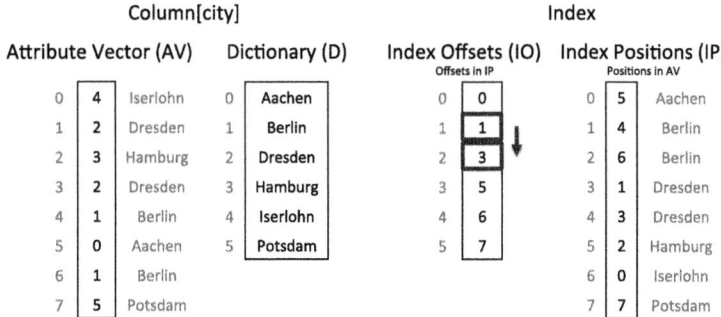

Fig. 18.6 Query processing using indices: Step 3

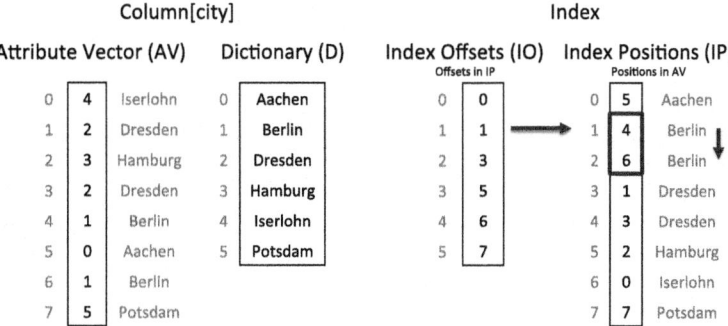

Fig. 18.7 Query processing using indices: Step 4

we are able to determine the exact range of all values we need to read in order to resolve the respective attribute vector positions of *Berlin*.

5. With the positions resolved in Step 4, we are able to jump directly to the respective attribute vector positions of all other columns of that table in order to materialize the complete records of all the people that live in Berlin (Fig. 18.8).

Using this approach, we reduce the data volume read by a CPU from the main memory by providing a data structure that does not require the scan of the entire attribute vector. Investigations regarding the influence of using indices on memory traffic and performance are shown in Sect. 18.4.

18.4   Discussion

In the previous section, we explained the idea of using an inverted index on a dictionary-encoded column to improve response time for lookup requests. An index increases the memory consumption per column. In this section we compare data

Column[city]   Index

Attribute Vector (AV)   Dictionary (D)   Index Offsets (IO)   Index Positions (IP)
                                              Offsets in IP   Positions in AV

AV		D		IO		IP	
0	4	Iserlohn	0	Aachen	0	0	0	5	Aachen
1	2	Dresden	1	Berlin	1	1	1	4	Berlin
2	3	Hamburg	2	Dresden	2	3	2	6	Berlin
3	2	Dresden	3	Hamburg	3	5	3	1	Dresden
4	1	Berlin	4	Iserlohn	4	6	4	3	Dresden
5	0	Aachen	5	Potsdam	5	7	5	2	Hamburg
6	1	Berlin					6	0	Iserlohn
7	5	Potsdam					7	7	Potsdam

Fig. 18.8 Query processing using indices: Step 5

Table 18.1 Symbols used for calculating memory consumption

Description	Unit	Symbol
Memory consumption of the index	bits	I_m
Length of the index offset vector	–	IO_l
Width of the index offset vector	bits	IO_w
Memory consumption of index offset vector	bits	IO_m
Length of the index positions vector	–	IP_l
Width of the index positions vector	bits	IP_w
Memory consumption of index positions vector	bits	IP_m
Length of dictionary (number of distinct values in column)	–	D_l
Length of attribute vector	–	AV_l
Width of attribute vector	bits	AV_w

lookup using full table scan against an index, regarding memory consumption and lookup performance. Table 18.1 introduces the symbols that we use.

18.4.1   Memory Consumption

In the beginning of this chapter, we explained that an index consists of an IO vector and an IP vector. To determine the overall size of the index, we need to calculate the size of these two structures.

$$I_m = IO_m + IP_m$$

The allocated memory of a vector can simply be calculated by multiplying its length (number of entries) with its width (size of a single entry).

$$IO_m = IO_l \cdot IO_w$$
$$IP_m = IP_l \cdot IP_w$$

The length of the index position vector *IP* directly corresponds to the length of the attribute vector AV_l, since it is basically a sorted version of the corresponding attribute vector. The width of IP is determined by the bit-encoded length of the attribute vector, since it contains direct positions to the values in the attribute vector.

$$IP_l = AV_l$$

$$IP_w = \lceil log_2(AV_l) \rceil \text{ bits}$$

The length of the index offset vector *IO* directly corresponds to the length of the dictionary D_l, which in turn is determined by the number of distinct values in the respective column. The width of *IO* is derived from the biggest offset into *IP*, because IO contains the bit-encoded offsets used to determine the position ranges in *IP*. As the maximum offset stored in IO can be the length of *IP*, the resulting width of *IO* is $\lceil log_2(IP_l) \rceil$ bits.

$$IO_l = D_l$$

$$IO_w = \lceil log_2(IP_l) \rceil \text{ bits}$$

Summarizing, we combine the mentioned formulas to a single equation for calculating the size of an index structure.

$$I_m = D_l \cdot \lceil log_2(IP_l) \rceil + AV_l \cdot \lceil log_2(AV_l) \rceil \text{ bits}$$

$$I_m = (D_l + AV_l) \cdot (\lceil log_2(AV_l) \rceil) \text{ bits}$$

Let us now calculate the actual size of an index for the *city* column of our *world_population* table from Fig. 18.1. We need to determine D_l, IP_l, and AV_l. Based on the assumption that there are about 1 million cities around the world and that the world population is 8 billion, we just need to insert these numbers into our formula.

$$I_m = 10^6 \cdot \lceil log_2(8 \cdot 10^9) \rceil + 8 \cdot 10^9 \cdot \lceil log_2(8 \cdot 10^9) \rceil \text{ bits}$$

So from this formula, we get an index size of about 31 GB for the *city* column.

18.4.2 Lookup Performance

Independent of using an index or not, we need to perform a binary search on the dictionary to determine the encoded value for the respective search term. Let us assume that we need to read $log_2(D_l)$ entries to perform the binary search. Since the binary search on the dictionary has to be done for both access methods we can ignore it, when we compare them.

In case of a full column scan, we need to traverse the attribute vector sequentially, by reading AV_l entries, each with a size of $\lceil log_2(D_l) \rceil$ bits. Again, assuming 8 billion rows in the table and 1 million cities, we need to read

$$8 \cdot 10^9 \cdot \lceil log_2(10^6) \rceil \text{ bits} = 160.000.000.000 \text{ bits}$$

for a full attribute vector scan.

Now, when using an index the situation is different. After the dictionary lookup, we directly read the upper and lower limit from the index offset vector (see Fig. 18.6). We neglect this step for our performance consideration since this constant effort does not impact the overall performance. Having determined the upper and lower limit for the index positions vector, we need to traverse through it (see Fig. 18.7). The number of entries to read from IP depends on the distribution of values in the column. Reading an attribute value that is used more frequently, we need to read more entries, on less frequently used values we need to read less. Assuming a uniform distribution of values we need to read $AV_l \div D_l$ entries. The width of the entries is $\lceil log_2(AV_l) \rceil$ bits. Combining both equations, we get

$$IndexPositions = \frac{AV_l \cdot \lceil log_2(AV_l) \rceil}{D_l}$$

for the number of bits to read from the index positions vector. Taking our world population example, where we look for all people living in Berlin, we come up with

$$\frac{8 \cdot 10^9 \cdot \lceil log_2(8 \cdot 10^9) \rceil \text{ bits}}{10^6} = 264.000 \text{ bits}$$

to read using an index. Assuming a CPU scan speed of 4 MB/ms/core, a single core needs about 5 s to scan the complete attribute vector. Accessing the column using an index, the CPU needs about 0.008 ms to read the attribute vector positions from IP of the people living in Berlin. Thus, in this example, we can improve the lookup performance drastically with the index, compared to a sequential attribute vector scan.

We compare the theoretical memory traffic for the attribute vector scan and the position read in Fig. 18.9 for different dictionary sizes on a column with 30 million entries. With a uniform distribution, the index leads to less memory traffic, if at least 8 distinct values are present.

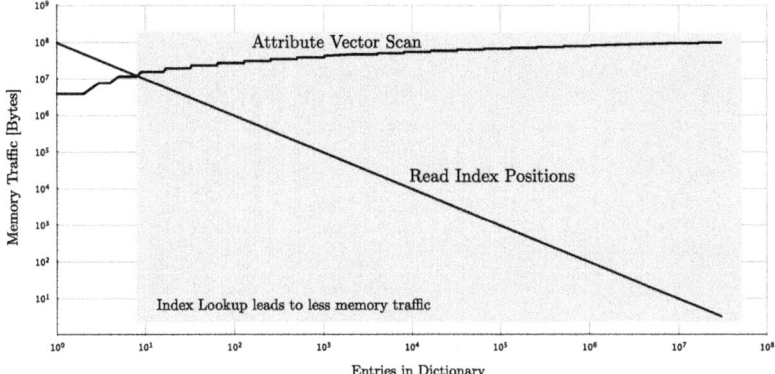

Fig. 18.9 Attribute vector scan vs. index position list read for a column with 30 million entries (note the log-log scale)

18.5   Self Test Questions

1. Index Characteristics
Introducing an index. . .

(a) increases memory consumption
(b) speeds up inserts
(c) slows down look-ups
(d) decreases memory consumption

2. Inverted Index
What is an inverted index?

(a) A list of text entries that have to be decrypted. It is used for enhanced security.
(b) A structure that maps each distinct value to a position list, which contains all positions where the value can be found in the column
(c) A structure that contains the distinct values of the dictionary in reverse order
(d) A structure that contains the delta of each entry in comparison to the largest value

Reference

[FSKP12]   M. Faust, D. Schwalb, J. Krueger, H. Plattner, Fast lookups for in-memory column stores: group-key indices, lookup and maintenance, in *ADMS '12: Proceedings of the 3rd International Workshop on Accelerating Data Management Systems Using Modern Processor and Storage Architectures at VLDB'12*, 2012

Chapter 19
Join

This chapter discusses the join operation and its execution in column-oriented main memory database systems. In general, joins are a way to combine tuples of two or more tables. There are two general categories of joins which can each be further specialized:

- *Inner Joins* create a result table that combines tuples from the two input tables based on a join predicate. It therefore combines each tuple from the first table with each tuple of the second table to evaluate the join predicate.
- *Outer Joins* are used to fetch information even though one relation does not have corresponding entries. An example where an outer join can be used is revenue (sum of sales order items) in a certain region and period. If the objective is to calculate the revenue for all regions over the whole year and there is a region with no revenue in a certain period then that information would be lost with a regular inner join. In contrast, the outer join inserts *Null* as the value for each attribute of the other relation, which does not yield a corresponding entry, into the result set.

Amongst others, we like to point out two further join specializations that are used in the examples of this section:

- *Equi Joins* are the most common join type. They allow the selection of tuples from both relations which satisfy a given equality predicate. Non-equi joins (e.g., join predicates using '<' or '>') are rather rare in enterprise systems.
- *Semi Joins* return only the matching tuples of the join result from one of the input relations. Tuples in the join result of the other relation are discarded [BC81].

The most prominent use case for joins are normalized database schemas where joins are executed based on foreign key relations.

There are different types of relationships between two tables. These are one-to-one, one-to-many, and many-to-many relationships. A one-to-one relationship connects one tuple of the first table with one tuple from the second table. This means, joining the tables yields exactly one or none tuple from one table for a tuple of the other table. In a one-to-many relationship, each tuple of the first

H. Plattner, *A Course in In-Memory Data Management*,
DOI 10.1007/978-3-642-55270-0_19, © Springer-Verlag Berlin Heidelberg 2014

table may be joined with multiple tuples of the second table. An example for a one-to-many relationship would be a normalized *world_population* table where a foreign key stores the country code (e.g., 'US' for the United States) and the full names of the countries are located in a separate *country* table that contains exactly one tuple per country. The resulting relationship between the tables *country* and *world_population* connects each country of the *country* table with all its citizens in the *world_population* table. In a many-to-many relationship each tuple of the first table is joined to multiple tuples from the second table and each tuple from the second table may be joined to multiple tuples from the first table. An example for a many-to-many relationship is the books-and-authors relationship in which a book can have many authors and an author can have multiple books. The many-to-many relationship is usually implemented as an additional table containing pairs of foreign keys that list the matching tuples of the two tables.

19.1   Join Execution in Main Memory

Looking at the properties of main memory shows that sequential scans are significantly faster than random accesses (see Chap. 14). Consequently, one target of join algorithms in main memory systems is to leverage sequential scans as much as possible. Therefore, random lookups should be avoided. Another target of these algorithms is to defer materialization of data as long as possible in order to work with the smaller positional information, e.g., to fit more data into one cache line (see Chap. 16).

The next two sections present the *hash join* and the *sort-merge join* as two join algorithms that leverage the features of in-memory databases when carefully adopted. The hash join uses a hash function to create a table with constant access times (i.e., a hash map) for the join predicates of the first relation. This hash map is then used in the second phase of the algorithm to probe each join predicate of the second relation for potential equality with values of the first relation. The sort-merge join works differently. Here, both relations are first sorted by the join predicate. The merge phase - which is the actual joining - can then be executed efficiently with sorted runs.

To discuss both algorithms, we use the *world_population* table and an additional *locations* table that stores details about specific locations (both simplified for the examples). See Fig. 19.1 for an overview of the example tables.

We furthermore use the statement in Listing 19.1. It retrieves the name of each city in the state 'Hessen' and calculates the corresponding population density, which is derived from dividing the number of inhabitants by the area of the city. For simplicity, we assume that city names are unique per state. As it is an inner join, only cities with at least one join partner in the *world_population* table are part of the result set.

world_population

recID	fname	lname	city	...
1	John	Doe	Berlin	...
2	Manfred	Mueller	Berlin	...
...	...	...	...	...
23	Max	Mustermann	Kassel	...
24	Hans	Gerhardt	Fulda	...

city dictionary of world_population

valueID	city
0	Aachen
1	Berlin
...	...
471	Kassel

locations

recID	city	state	country	area
1	Berlin	Berlin	Germany	851
2	Potsdam	Brandenburg	Germany	187
...	...	...	...	...
902	Kassel	Hessen	Germany	107

city dictionary of locations

valueID	city
0	Aachen
1	Berlin
...	...
81	Kassel

Fig. 19.1 The example join tables

```
SELECT wp.city, (COUNT(*) / locations.area) AS population_density
FROM world_population AS wp
INNER JOIN locations ON wp.city = locations. city
WHERE locations.state = 'Hessen'
GROUP BY wp.city, locations.area
```

Listing 19.1 SQL statement of an inner join between *world_population* and *locations*

19.2 Hash Join

The hash join is based on a hash function that provides access to items in the joining data structure in constant time. A hash function maps arbitrary inputs to fixed length keys, even though the inputs might have variable lengths. The joining data structure for the hash join is a so-called hash map, which implements an associative array that maps keys to values.

The hash join algorithm itself consists of two phases: a hash phase (also called build phase) and a probe phase. During the hash phase, the smaller relation (i.e., the relation with the lower table cardinality) is sequentially scanned on the join-predicate column and the hash key (let us assume an integer value for simplification) of each attribute value is calculated. The recordID of the current tuple is inserted into the hash map at the position that is determined by the hash key. To keep the hash map small and its creation fast, the smaller one of the two relations to join is used for the hash map creation resulting in improved cache utilization.

During the probe phase, the larger relation is sequentially scanned. Each value is probed into the hash map by calculating the hash key and looking it up in the hash map. If the value exists, the position of the currently probed tuple and the other

hash value	positions in locations
hash(81)	902
hash(37)	912
hash(42)	1023
...	...

Fig. 19.2 Hash map for *locations* table created in hash phase

tuples' positions from the hash map are returned as matching pairs. A row is skipped if its key does not exist in the hashmap.

19.3   Example Hash Join

This section provides a deeper insight into the hash join algorithm by analyzing the example query's execution as displayed in Listing 19.1. The first step is to find all predicates that can be evaluated before the actual join execution starts. This allows reducing the size of intermediate results and in turn saves memory and bandwidth. Regarding the example, this means the filter operation of the *WHERE* clause is executed beforehand. The result is a reduced amount of cities that are relevant for the join. In our example, the result is a list of all positions of cities of the *locations* table in 'Hessen', which are e.g. [902, 912, 1023, . . .] for ['Kassel', 'Darmstadt', 'Fulda', . . .].

At this point the join operation starts with the hash phase. The cities in 'Hessen' in the *locations* table are the smaller input relation, which means they are sequentially scanned to build the hash map as shown in Fig. 19.2. The hash map is created by hash-encoding the valueID of the scanned line item as the key. Depending on the input data, also the valueID can be used directly as an offset without additional hashing. The value for each key is a list of positions (i.e., the recordIDs) where the valueID is found. In our example each of these lists store only one value, since the *city* column of the *locations* table is distinct. In the case of 'Kassel' with valueID 81 that is 902 while the positional list for 'Darmstadt' with the valueID 37 stores the position 912 (see Fig. 19.2).

Because SanssouciDB uses dictionary encoding, an additional step has to be executed in comparison to the traditional hash join. The reason is that the values of the *city* columns of both tables cannot be compared directly since they both have their own dictionaries with potentially different valueIDs. Therefore a mapping is created that maps each valueID of the *city* column in the *world_population* table to its corresponding valueID of the *city* column in the *locations* table as shown in Fig. 19.3.

The next phase is the actual joining of both tables, which is called *probe phase*. Figure 19.4 displays one step of the example's probe phase. The *city* column of the

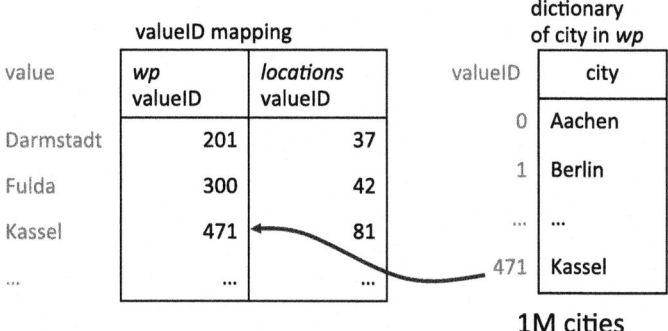

Fig. 19.3 Mapping of ValueIDs between tables *world_population* and *locations*

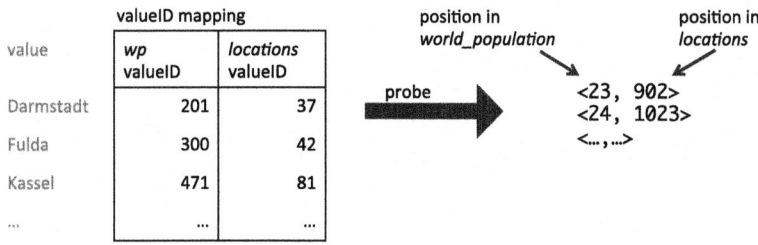

Fig. 19.4 Probe phase

world_population table is scanned sequentially. For each valueID of that column, the corresponding valueID for the *city* column of the *locations* table is looked up in the mapping structure. The corresponding valueID is then probed into the hash map to check if the hash map contains the key. This access is of constant complexity. When a key is found, the recordID of the currently probed row and the recordIDs that are stored together with the matching hash key are returned as matching pairs. The resulting pairs are the result of the actual join operation.

Afterwards, the remaining steps of the query are executed. This includes executing the COUNT aggregation based on the found position pairs to calculate how many people are living in each city. Finally, the name of the city is fetched and the population density is calculated to complete the query result.

A significant challenge when implementing hash joins is to choose a hash function that hashes each value efficiently and with as few collisions as possible. Collisions occur when the hashes of two distinct values are equal. Consequently the hash join has to deal with collisions, which makes it more complex. An alternative to the hash join is the sort-merge join, which will be discussed in the next section.

Please note that due to the nature of hashing the hash join algorithm is only applicable for equi joins.

19.4 Sort-Merge Join

The sort-merge join consists of two phases. In the first phase, the sort-merge join scans the join predicate of the smaller relation and creates a list of pairs, where each pair consists of valueID and recordID, and sorts this list by valueID (we refer to that list of pairs as A). In parallel, a list of unique valueIDs is created (referred to as U).

Using U a mapping structure similar to the one described in the hash join section is created. This mapping structure is needed to ensure that the valueIDs from both tables refer to the same value in their corresponding dictionaries. With the help of the mapping structure, each valueID in A is replaced with the corresponding valueID of the larger relationship.

In the second phase, the larger relation is sequentially scanned and a list of pairs called B is created that has a similar structure as A. Afterwards B is being sorted by valueID. Now we have two lists of pairs, i.e. A and B, that use the same value encoding and are both sorted by valueID. Consequently merging can be executed efficiently as shown in more detail with a comparable algorithm in Sect. 27.2. While merging, the join emits a list of position pairs with matching tuples that are afterwards materialized and returned.

19.5 Example Sort-Merge Join

This section explains the execution of the sort-merge join in more detail, using the query from Listing 19.1 again. To show how unmatched join predicates are handled, we assume that the *world_population* table does not contain any inhabitants of 'Darmstadt'. An overview of both join tables and their join column dictionaries is shown in Fig. 19.1. 'Kassel', the city of the last example, has position 902 in the *locations* table. The valueID of 'Kassel' is 471 in *world_population*'s *city* dictionary and 81 in *locations*'s *city* dictionary.

The first phase of the sort-merge join execution is similar to the execution of the query using the hash join algorithm: The filter operation is executed to reduce the number of input tuples before the actual join operation. With the remaining tuples in the *locations* table, list A is created containing pairs consisting of valueID and recordID, e.g., *<81, 902>* for 'Kassel' and *<37, 912>* for 'Darmstadt'. This list of pairs is then sorted by valueID.

Next, a unique list of valueIDs is created using A. Consequently, this list contains all valueIDs of the cities in 'Hessen', i.e. *[902, 912, 1023, …]*. Afterwards the list is sorted and is used to create the mapping structure. The structure maps between valueIDs of the *city* dictionaries of *world_population* and *locations*. As shown in Fig. 19.5, valueID *37* from the *locations* table does not have a corresponding valueID in the *world_population* table. Hence, there are no join partners for that value and consequently the entry is removed from the mapping structure as well as

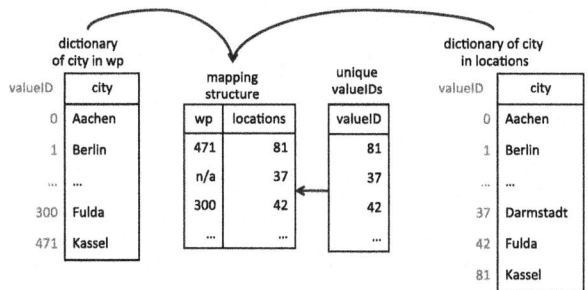

Fig. 19.5 Building the mapping structure

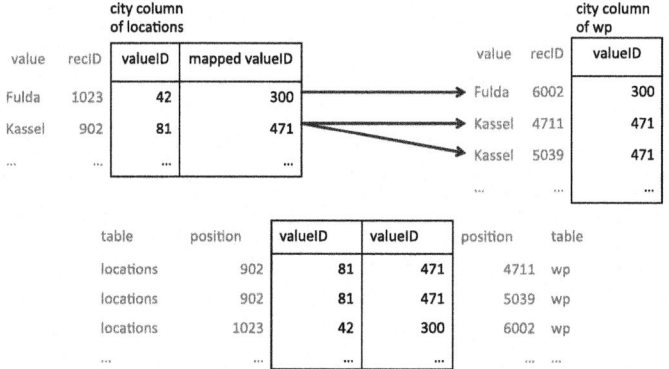

Fig. 19.6 Matching pairs from both position lists

from list A. Since A is sorted by valueID, the lookups to the mapping structure are sequential and consequently leverage pre-fetching and cache utilization.

The next step is to translate the valueIDs in list A with their corresponding valueIDs in *world_population*'s city column. This yields a list of tuples that changed from the form of $<valueID_{location}, recordID>$ to $<valueID_{world\_population}, recordID>$. In our example, the corresponding tuple for 'Kassel' would change from $<81, 902>$ to $<471, 902>$. Since it is safe to assume that both dictionaries for the city columns had the same sorting order, an additional sorting after the valueID translations is not necessary.

The second phase of the sort-merge join creates list B, which has a similar structure as A and is being sorted by valueID, too. After B is finalized, the two lists of the *world_population* and *locations* tables are sorted by valueID and use the same value encoding (i.e., valueIDs of *world_population*). That means it is now possible to merge both lists sequentially and create a list of matching pairs in the form of $<recordID_{location}, recordID_{world\_population}>$. Please note that the merging in this case has to deal with non-distinct lists so the cross product of all matching valueIDs has to be built. This step is illustrated in Fig. 19.6.

The last step of the sort-merge join is similar to the last step of the hash join. The aggregation, i.e., the number of inhabitants for each city in 'Hessen', is executed for the calculation of the population density in order to create the result set.

19.6   Choosing a Join Algorithm

Which join algorithm to chose for optimal performance depends on several factors. Amongst them are the size of the relations to join, the algorithm's complexity, and constraints given by the current workload. In this section we discuss the complexity of the presented join algorithms. For simplification we ignore the additional work that has to be done to evaluate predicates on columns using different value encodings, which includes the mapping structure creation and the lookups into that structure. Symbols m and n denote the number of entries of the two tables that are being joined.

The hash join has a complexity of $\mathcal{O}(n + m)$ as the first relation's join column is scanned once to build the hash map and afterwards the second relation's join column is scanned once to probe the values into the hash map.

The complexity of the sort-merge join is $\mathcal{O}(n \cdot log(n) + m \cdot log(m))$. In general, it is worse than the hash join's complexity because it requires additional sorting before the merge phase.

Besides hash join and sort-merge join, the *nested-loop join* is another option to perform a join. Basically, it scans a relation and for each scanned tuple, the whole other relation is scanned for matching tuples. This results in a complexity of $\mathcal{O}(n \cdot m)$.

To sum up, the hash join performs best in general. The limitation of the algorithm is the hash map. If the map becomes too large or it is too complicated to build at all, there might be better alternatives. Albutiu et al. [AKN12] has shown that highly optimized sort-merge joins adapted for NUMA architectures and multi-core processing are amongst the fastest join algorithms as of now. The nested-loop algorithm is suitable for very small data sets, where creating the data structures of the other algorithms would create too much overhead compared to the actual joining costs.

19.7   Self Test Questions

1. **Sort-Merge Join Complexity**
   What is the complexity of the Sort-Merge Join?

   (a) O(n · m)
   (b) O(n²/m²)
   (c) O(n+m)
   (d) O(n · log(n)+m · log(m))

2. **Join Algorithm for Small Data Sets**
   Given is an extremely small data set. Which join algorithm would you choose in order to get the best performance?

   (a) Nested-Loop Join
   (b) All join algorithms have the same performance
   (c) Hash Join
   (d) Sort-Merge Join

3. **Equi Join**
   What is the Equi Join?

   (a) It is a join algorithm to fetch information, that is probably not there. So if you select a tuple from one relation and this tuple has no matching tuple on the other relation, you would insert NULL values there.
   (b) It is a join algorithm that ensures that the result consists of equal amounts from both joined relations
   (c) If you select tuples from both relations, you use only one half of the join relations and the other half of the table is discarded
   (d) If you select tuples from both relations, you will always select those tuples, that qualify according to a given equality predicate

4. **One-to-One-Relationship**
   What is a one-to-one relationship?

   (a) Each query which has exactly one join between exactly two tables is called a one-to-one relationship, because one table is joined to exactly one other table.
   (b) A one-to-one relationship between two objects means that for exactly one object on the left side of the join exists exactly zero or one object on the right side and vice versa
   (c) A one-to-one relationship between two objects means that for each object on the left side, there are one or more objects on the right side of the joined table and each object of the right side has exactly one join partner on the left
   (d) A one-to-one relationship between two objects means that each object on the left side is joined to one or more objects on the right side of the table and vice versa each object on the right side has one or more join partners on the left side of the table

5. **Join Algorithms**
   Which of these Join algorithms exists?

   (a) Reverse-Traversal Join
   (b) Bubble Join
   (c) Bootstrap Join
   (d) Nested-Loop Join

References

[AKN12]   M.-C. Albutiu, A. Kemper, T. Neumann, Massively parallel sort-merge joins in main memory multi-core database systems. PVLDB **5**(10), 1064–1075 (2012)

[BC81]   P.A. Bernstein, D.-M.W. Chiu, Using semi-joins to solve relational queries. J. ACM **28**(1), 25–40 (1981)

Chapter 20
Aggregate Functions

This chapter discusses aggregate functions. It outlines what aggregate functions are, how they work, and how they can be executed in an in-memory database system.

Aggregate functions are specific functions that take multiple rows as an input to create an output. This means, they work on data sets instead of single values. Grouping the input data based on specified group attributes creates the sets. Basic aggregate functions are, e.g., *COUNT*, *SUM*, *AVERAGE*, *MEDIAN*, *MAXIMUM* and *MINIMUM*. Furthermore, it is possible to create additional functions for special purposes, e.g., OLAP functions that extend basic functions. The basic SQL syntax to use an aggregate function can be seen in Listing 20.1.

The *GROUP BY* clause specifies the attributes by which the input relation is grouped. All selected attributes that are not part of the *GROUP BY* clause should specify an aggregate function in the select clause, otherwise their value might be undefined. The *WHERE* and the *HAVING* clauses are optional.

20.1 Aggregation Example Using the COUNT Function

Let us consider an example for the use of the *COUNT* aggregate function. Assume an input table containing the complete world population as shown in Fig. 20.1.

The goal is to count all citizens per country. Using the *COUNT* aggregate function, an SQL query to achieve this is depicted in Listing 20.2:

Figure 20.2 shows how such a query would be processed. First, the system runs through the attribute vector for the *country* column. For each new encountered country valueID, an entry with initial value "1" is added to the result map. If the encountered valueID has been added before, the respective entry in the result map is increased by one. That way, a result map is created which contains the valueIDs of each country and its number of occurrence. Second, the actual country names are fetched from the country dictionary using the valueIDs and the final result of

```
SELECT AggregateFunction(attribute1), attribute2, attribute3
FROM table_name
WHERE attribute2 = some_value
GROUP BY attribute2, attribute3
HAVING AggregateFunction(attribute1) > 5;
```

Listing 20.1 SQL aggregate function syntax

recID	fname	lname	gender	city	country	birthday
0	John	Smith	m	Chicago	USA	12.03.1964
1	Mary	Brown	f	London	UK	12.05.1964
2	Jane	Doe	f	Palo Alto	USA	23.04.1976
3	John	Doe	m	Palo Alto	USA	17.06.1952
4	Peter	Schmidt	m	Potsdam	GER	11.11.1975
...	...	...	...		...	...

Fig. 20.1 Input relation containing the world population

```
SELECT country, COUNT(*) AS citizens
FROM world_population
GROUP BY country;
```

Listing 20.2 Example SQL query using the COUNT aggregate function

Attribute Vector for "country"

44	43	44	44	42	...

Result Map

44	3
43	1
42	1
...	...

\+

country Dictionary

valueID	value
...	...
42	GER
43	UK
44	USA
...	...

=

Result

country	citizens
USA	3
UK	1
GER	1
...	...

Fig. 20.2 Count example

the *COUNT* function is created. The result contains pairs of country names and the countries' number of occurrences in the source table, which corresponds to the number of citizens.

20.2   Other Aggregate Functions

Other aggregate functions require a similar calculation function. For *SUM*, the number of occurrences for each valueID is counted in an auxiliary data structure and the sum is calculated in a final step by summing up the number of occurrences multiplied with the respective value of each valueID. *AVERAGE* can be calculated by dividing *SUM* by *COUNT*. To retrieve the median, the complete relation has to be sorted and the middle value is returned. *MAXIMUM* and *MINIMUM* compare a temporary extreme value with the next value from the relation and replace it if the new value is higher (for *MAXIMUM*) or lower (for *MINIMUM*), respectively.

20.3   Self Test Questions

1. **Aggregate Function Definition**
   What are aggregate functions?

   (a) A set of indexes that speed up processing a specific report
   (b) A specific set of functions that summarize multiple rows from an input data set
   (c) A set of functions that transform data types from one to another data type
   (d) A set of tuples that are grouped together according to specific requirements

2. **Aggregate Functions**
   Which of the following is an aggregate function?

   (a) GROUP BY
   (b) MINIMUM
   (c) HAVING
   (d) SORT

Chapter 21
Parallel Select

21.1 Context

Chapter 18 introduced the concept of an inverted index to prevent the database system from performing a full column scan every time a query searches for a predicate in the column. However, maintaining an index consumes additional memory and comes with additional processing overhead. So the decision to use an index should be made carefully, balancing all the pros and cons an index would bring in the particular situation. This chapter discusses how to speed up a full column scan despite adding an index to it. Chapter 17 introduced parallelism as an action to parallelize the execution of database operations. In this chapter, we present a detailed description of how parallelism can be used to speed up the execution of a SELECT operation.

21.2 Parallelization Example

The purpose of a SELECT operation is to find and return all positions of values in a column that fulfill the SELECT predicate. Thus, we need to scan the attribute vector to find the position of all valueIDs matching the dictionary entry of the predicate. Splitting the vector into chunks of data enables parallel execution of a sequential scan across the full attribute vector of a column. Thus, each chunk of data can be processed independently while the results of all subsequent scan operations are combined.

Consider the following example. We want to find the names of all male persons from Italy. The corresponding database query to answer this question is depicted in Listing 21.1. Please note that this query was simplified for better understanding of the concepts and does not reflect the optimal query.

H. Plattner, *A Course in In-Memory Data Management,*
DOI 10.1007/978-3-642-55270-0_21, © Springer-Verlag Berlin Heidelberg 2014

SELECT fname , lname
FROM world_population
WHERE country = 'IT' **AND** gender = 'm';

Listing 21.1 Query to select all male persons from Italy

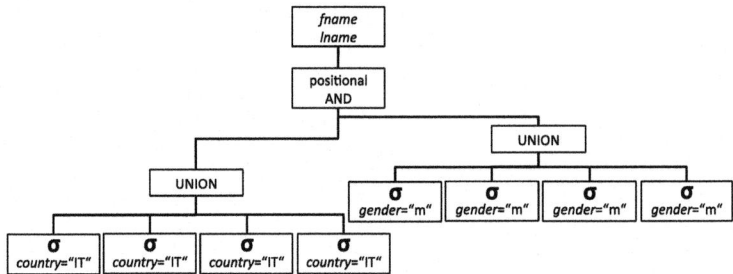

Fig. 21.1 Parallel predicate scan on partitioned attribute data

Figure 21.1 shows the resulting query plan when columns are split into four data partitions. In this case, the scan of the attribute vector can be performed in parallel by eight independent threads, i.e. four for each of the predicate attributes country and gender.

As can be seen in Fig. 21.1, two *UNION* operations are needed to combine the intermediate results obtained by the individual threads while the *positional AND* operation is executed sequentially. Figure 21.2 shows such a scenario based on our example. The two attribute vectors of the attributes country and gender are partitioned into four equally-sized chunks. The position ranges in the respective chunks are the same for both columns (0...2, 3...5, 6...8, and 9...11). Parallel predicate scans in the individual chunks will result in the selection shown in Fig. 21.3. After all parallel predicate scans determined the individual positions that correspond to the search predicate in the respective column, we need to compare the positions within each chunk. If the same position has been marked in both columns, we have found a record that fulfills the complete search predicates.

Figure 21.4 shows the result of the comparison.

The last operation in the query execution is the *UNION*, which combines the results from the intermediate *positional AND* operations.

Fortunately, the *positional AND* operation can be parallelized as well. If both attribute columns are equally partitioned, meaning that the number of partitions is equal and the position ranges in each corresponding partition are the same, the *positional AND* can be executed in parallel as well. Once, the *positional AND* is parallelized the executed query plan changes. The new query plan is depicted in Fig. 21.5. It differs from the one shown in Fig. 21.1 in that way that the positional AND operation is performed on each of the intermediate results requiring only a single UNION operation at the end.

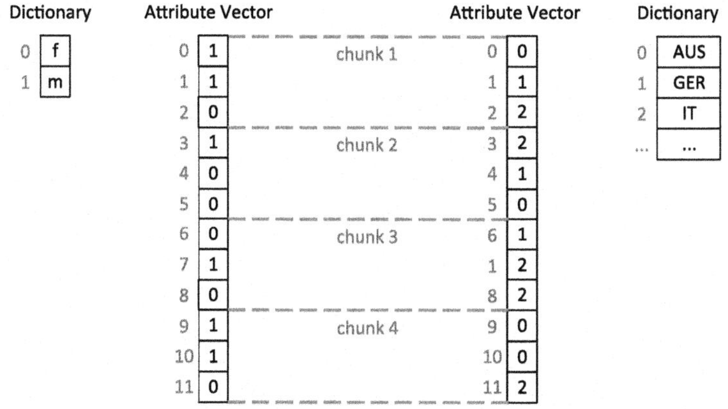

Fig. 21.2 Equally partitioned columns

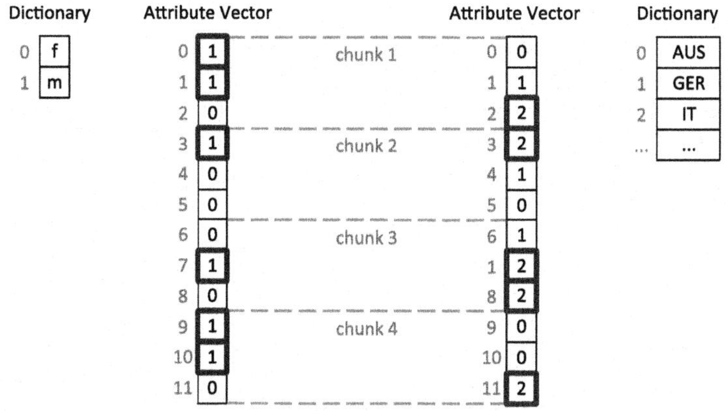

Fig. 21.3 Result of parallel predicate scans

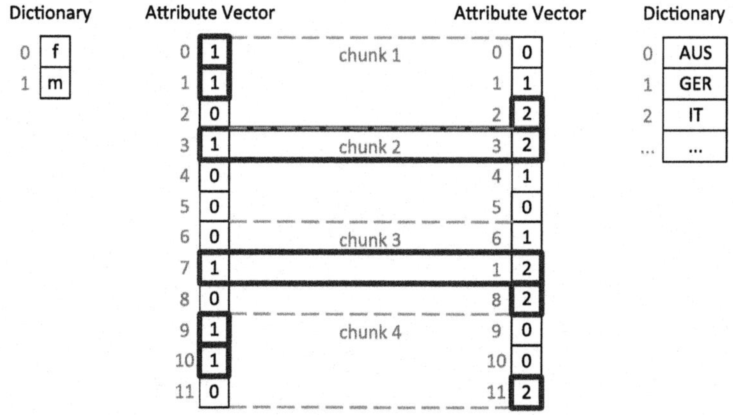

Fig. 21.4 Result of *positional AND*

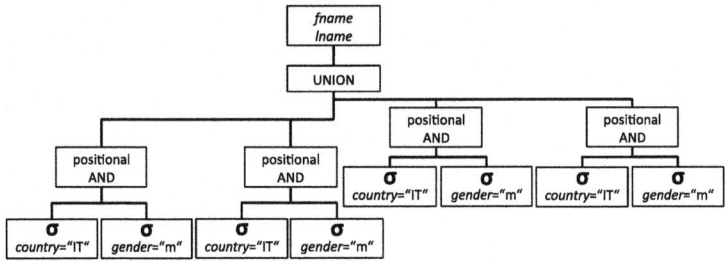

Fig. 21.5 Parallel predicate scans with parallel *positional AND* operation

21.3   Self Test Questions

1. **Query Execution Plans in Parallelizing SELECTS**
   When a SELECT statement should be executed in parallel . . .

   (a) all other SELECT statements have to be paused
   (b) its query execution plan is not changed at all
   (c) its query execution plan becomes much simpler compared to sequential execution
   (d) its query execution plan has to be adapted accordingly

Chapter 22
Workload Management and Scheduling

22.1 The Importance of Speed

One of the most important factors that determine the usability for interactive applications is response time. Psychological studies show that the acceptable maximum application response time for a human is about 3 s. After 3 s, the user might loose concentration and do something else. Once the application has finished its processing and is ready for the next human input, the user has to refocus on the application again. These context switches are extremely expensive as they constantly interrupt the user and lead to being unproductive. Consequently, an important objective for data management is to answer queries issued by interactive applications fast.

As described in Chap. 3, it is commonly distinguished between analytical and transactional applications. Both types of applications issue a specific workload to the underlying database. Although not formally defined, it is widely agreed that transactional queries are considered as short-running, whereas analytical queries are typically being classified as long-running. Due to their specific fields of application, they also differ in their service level objectives: Transactional applications are often coupled to customer interactions or other time-critical business processes and have high response time requirements. On the other hand, analytical applications that generate reports are typically considered batch-oriented with comparably less critical response-time requirements. As introduced in Chap. 2, new requirements in enterprise computing demand for fast execution of complex, analytical queries. A table scan necessary for an OLAP query can easily be parallelized to a high degree by a query optimizer. If the query optimizer decides that the query can be executed in parallel, it separates the query into multiple sub-queries, executes every sub-query in parallel and combines their intermediate results, resulting in a reduced overall execution time. This parallel execution is required to reach response times of merely seconds, as expected by interactive users.

Obviously, processing a mixed workload of response time critical transactions, as well as complex analytical queries can potentially lead to resource contention in the database system. Based on the query classes discussed above, we can state a number of desirable design objectives for a database system executing a mix of these query classes. First, queries need to run concurrently to efficiently use available resources: as transactional queries typically run sequentially, allowing only one query at a time would result in an underutilization of the system. Also, analytical queries or other database tasks will have parts that cannot be parallelized to a degree that leverages all available processing units. Second, transactional queries need to run with a higher priority over other query classes: given the relatively short run-time of transactional queries compared to the other classes, they cannot wait until queries of the other classes have completed to meet their service level objectives. Third, interactive analytical queries have to gain access to a large number of resources quickly: given the tight response time requirements and the high complexity, queries of this type need to be executed with a high degree of intra-query parallelism and therefore require access to a larger number of available processing units. And fourth, if a query is running on a number of processing units in parallel, it needs to free up resources quickly in case a query with higher priority arrives for execution. Consequently, some form of workload management is required in the database systems to enforce these requirements.

22.2　Scheduling

The database scheduler is the component that decides which of the incoming query or sub query gets executed on the available computing resources. In general, scheduling goes along with several kinds of complications: inter-dependencies between queries, different resource utilization for the queries (e.g. some are CPU-bound or memory-bandwidth bound), restricted resources, locality of operations (e.g. a filter task should be executed at the node where the data is stored) and so on. These possible complications make finding an optimal execution plan a very complex task which is NP-hard in most cases. The problem of finding an optimal scheduling becomes even more complex, if we consider the online arrival of multiple queries. However, approximate solutions often outperform complicated algorithms in practice.

To achieve the desired system characteristics as introduced in the previous section, the scheduler follows a task based execution model. The logical query plan of an incoming query is transformed into a set of atomic tasks that represent this plan. These tasks may have data dependencies, but can be executed otherwise independently, thus forming a directed acyclic graph. A task-based approach provides two main advantages over considering a whole query as the unit for scheduling: More fine granular units of scheduling allow for better load balancing on a multiprocessor system, as well as more control over progress of query execution based on priorities. Splitting queries into small units of work introduces points in time during query execution, where lower priority queries can be paused to run higher priority queries without the need of canceling or preempting the low priority query. A more detailed discussion of task based scheduling can be found in [WGP13].

22.3 Mixed Workload Management

Typically, optimizing resource usage and scheduling is a workload-dependent problem and becomes even more complicated when two different types of workloads are mixed. As said before, a transactional workload is characterized by short running queries that must be executed within tight time constraints. In contrast, an analytical workload consists of more complex and computationally heavier queries. Running mixed workloads on a single database instance leads to potentially conflicting optimization goals. The response time for transactional queries must be guaranteed. This can be achieved by assigning priorities to tasks corresponding to critical transactional queries and adopt a priority scheduling policy in the task scheduler. At the same time, the response time for analytical queries should be as short as possible. This is achieved by using the aforementioned task based scheduler, which makes sure that the available resources are fully utilized.

22.4 Self Test Questions

1. **Resource Conflicts**
   Which three hardware resources are usually taken into account by the scheduler in a distributed in-memory database setup?

   (a) Main memory, disk and tape drive sizes
   (b) CPU processing power, main memory size, network bandwidth
   (c) Network bandwidth, power supply unit, main memory
   (d) CPU processing power, graphics card performance, monitor resolution

2. **Workload Management Scheduling Strategy**
   Why does a complex workload scheduling strategy might have disadvantages in comparison to a simple resource allocation based on heuristics or a uniform distribution, e.g. Round Robin?

   (a) A scheduling strategy is based on general workloads and thus might not reach the best performance for specific workloads compared to heuristics or a uniform distribution, while its application is cheap.
   (b) Heuristics are always better than complex scheduling strategies.
   (c) The execution of a scheduling strategy itself consumes more resources than a simplistic scheduling approach. A strategy is usually optimized for a certain workload—if this workload changes abruptly, the scheduling strategy might perform worse than a uniform distribution.
   (d) Round-Robin is usually the best scheduling strategy.

3. **Analytical Queries in Workload Management**
   Analytical queries typically are ...

   (a) short running with soft time constraints
   (b) short running with strict time constraints
   (c) long running with soft time constraints
   (d) long running with strict time constraints

4. **Query Response Times**
   Query response times ...

   (a) have to be as short as possible, so the user stays focused at the task at hand
   (b) can be increased so the user can do as many tasks as possible in parallel because context switches are cheap
   (c) have no impact on a users work behavior
   (d) should never be decreased as users are unfamiliar with such system behavior and can become frustrated

Reference

[WGP13] J. Wust, M. Grund, H. Plattner, Tamex: a task-based query execution framework for mixed enterprise workloads on in-memory databases, in *GI-Jahrestagung*, pp. 487–501, 2013

Chapter 23
Parallel Join

The join operation is one of the most cost-intensive tasks for in-memory databases. As today's systems introduce more and more parallelism, intra-operator parallelization moves into focus. This chapter discusses possible schemes to parallelize the hash join algorithm, as described in Chap. 19. The hash join algorithm usually consists of two phases:

1. The hashing phase: creation of a hash map for the smaller join relation.
2. The probing phase: sequentially scanning the larger join relation while probing into the hash map created in phase one.

As discussed in Sect. 19.1, the performance of the hash join algorithm can be optimized by accessing memory sequentially and materializing as late as possible. Various methods exist to further improve performance and exploit multi-core architectures. We will first outline a simple, only partially parallelized hash join algorithm in Sect. 23.1, followed by a more complex and fully parallelized version in Sect. 23.2.

Because this chapter focusses on the parallelization of the hash join operation, we will not discuss the implications of joining two columns with different encodings (see Sect. 19.3 for more details). A and B denote the input relations of the join operation, where A is the smaller relation.

23.1 Partially Parallelized Hash Join

A simple way to parallelize a hash join is to keep the hashing phase sequential and to only parallelize the probing phase.

In the sequential hashing phase, a hash table for relation A is created. This is done by sequentially scanning the join column, computing the respective hash value of each element and storing the position in the element's hash bucket.

In the probing phase, relation B is partitioned horizontally across the available threads and probing is performed in parallel. Each thread works with a local copy

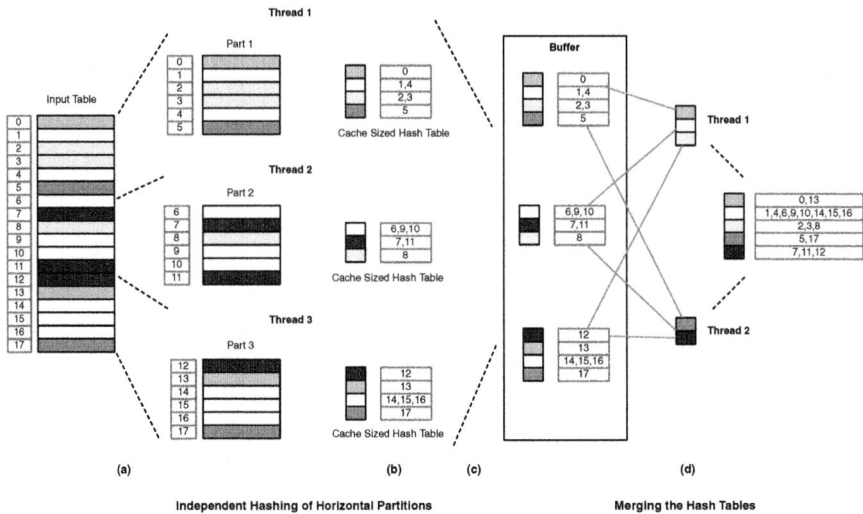

Fig. 23.1 Parallelized hashing phase of a join algorithm

of the hash table and stores the join results in a local result table. When all probe threads have finished, the local result tables are merged into a unified result table.

Although the join is not fully parallelized, this approach works well in practice, as the probing phase on the larger join relation B dominates the join costs.

23.2 Parallel Hash Join

A more complex parallelization approach executes both, the hash phase as well as the probing phase of the join, in parallel. While the second phase is similar to the approach described in Sect. 23.1, also the hash tables for the smaller input relation A are calculated in parallel, as outlined in Fig. 23.1.

First, the hash table for relation A is created. For that, multiple *hash threads* work independently on relation A, which is sliced up into multiple partitions as shown in Step (a) of Fig. 23.1. Each hash thread scans its part of the input table, hashes the values and writes its results into a small local hash table, as outlined by Step (b). When the local hash table of a thread reaches a predefined size limit (e.g., its size exceeds a certain cache level size), it is written back to a buffer and a new local hash table is created by the thread. Each local hash table is added to a queue, symbolized by the *buffer* in Step (c). Multiple *merge threads* process the added tables in the buffer, merging them to a unified hash table for relation A, depicted in Step (d). Each merge thread only processes its exclusively assigned values, so that the write synchronization on the unified hash table can be reduced to a minimum.

In the second phase, probing is parallelized over the available threads as outlined before in Sect. 23.1. This means, the larger join relation B is partitioned horizontally and each thread stores its results in a local result table. When all threads have finished, the local result tables of the probe threads are merged. For a more detailed discussion of parallel join algorithms, the interested reader is referred to [MBK82, KKL$^+$09, AKN12].

23.3 Self Test Questions

1. **Parallelizing Hash-Join Phases**
   What is the disadvantage when the probing phase of a join algorithm is parallelized and the hashing phase is performed sequentially?

   (a) The algorithm still has a large sequential part that limits its potential to scale
   (b) Sequentially performing the hashing phase introduces inconsistencies in the produced hash values
   (c) The table has to be split into smaller parts, so that every core, which performs the probing, can finish
   (d) The sequential hashing phase will run slower due to the large resource utilization of the parallel probing phase

References

[AKN12] M.-C. Albutiu, A. Kemper, T. Neumann, Massively parallel sort-merge joins in main memory multi-core database systems. Proc. VLDB **5**(10), 1064–1075 (2012)

[KKL$^+$09] C. Kim, T. Kaldewey, V.W. Lee, E. Sedlar, A.D. Nguyen, N. Satish, J. Chhugani, A. Di Blas, P. Dubey, Sort vs. hash revisited: fast join implementation on modern multi-core CPUs, in *VLDB*, 2009

[MBK82] S. Manegold, P. Boncz, M. Kersten, Optimizing main-memory join on modern hardware. IEEE Trans. Knowledge Data Eng. **14**, 709–730 (2002)

Chapter 24
Parallel Aggregation

Similar to the parallel join described in Chap. 23, aggregation operations can also be accelerated using parallelism and hash-based algorithms. In this chapter, we discuss how parallel aggregation is implemented in SanssouciDB. Note that multiple other ways to implement parallel aggregation are also conceivable. However, we focus on our parallel implementation using hashing and thread-local storage.

24.1   Aggregate Functions Revisited

The concept of aggregate functions has already been discussed in Chap. 20. Aggregate functions operate on single columns but usually take a large number of tuples into account. Examples for aggregate functions are *COUNT, SUM, AVERAGE, MIN, MAX* or *STDDEV*. Which aggregates are returned is typically specified using *GROUP BY* and/or *HAVING* clauses in SQL. The *GROUP BY* clause is used to express that the aggregate function shall be computed for every distinct value of the specified attribute(s). For example, the following query would incur a *COUNT* operation with two different results, namely one for female and another one for male entries:

```
SELECT gender , COUNT(*)
FROM world_population
GROUP BY gender ;
```

Listing 24.1   Simple SQL query using the COUNT aggregate function

The *HAVING* clause behaves similar to the SQL *WHERE* clause, the difference being that the filter criterion has to contain aggregate functions and is evaluated on the groups defined by the *GROUP BY* clause.

H. Plattner, *A Course in In-Memory Data Management*,
DOI 10.1007/978-3-642-55270-0_24, © Springer-Verlag Berlin Heidelberg 2014

24.2 Parallel Aggregation Using Hashing

Let us revisit the following example from Chap. 20:

```
SELECT country , COUNT(*) AS citizens
FROM world_population
GROUP BY country ;
```

Listing 24.2 SQL Example to count the citizens of each country

The result of this query lists the number of all citizens for every distinct value of the "country" column. Table 24.1 shows the result of the query in Listing 24.2.

In the following, we describe how this result is computed using the parallel aggregation algorithm in SanssouciDB. Figure 24.1 visualizes how the algorithm works. It consists of two phases, the hashing phase and the aggregation phase.

In the hashing phase, shown in the upper part of Fig. 24.1, the world population table is horizontally partitioned into n parts (cf. Sect. 9.3). The chosen n determines how many threads will be used for the parallel aggregation. In a lightly loaded system, n might be as high as the number of CPU cores available on the machine (cf. Chap. 22). Note that the horizontal partitioning occurs logically and dynamically, i.e. the way the table is stored physically remains unaltered. Each of the n threads is now assigned a partition and creates an empty thread-local hash table. The hash table is used to store

- the hash value of a country and
- the number of occurrences of that country in that partition of the world population table.

The threads then begin to scan the "country" column. For each row in the partition of a thread, the thread checks whether the current country is already contained in its thread-local hash table. If yes, the number stored in correspondence with the hash value of the country is increased by one. If not, the hash value of the country is stored along with a "1", since the thread has just found the first inhabitant of the country in the current partition. Note that creating a new entry in the hash table might result in a situation where the size of the hash table exceeds the size of the CPU cache. Whenever this would be the case, the hash table is stored in main memory and a new hash table is created. Ensuring that the hash table never exceeds the size of the CPU cache is crucial for facilitating fast lookups in the hash table. Since a lookup occurs for every row, which is being scanned, the lookup and alteration of the hash table should not incur a cache miss. When all threads have finished scanning their assigned partition, the aggregation phase begins.

In the aggregation phase, the buffered hash tables are merged. This is accomplished by so-called merger threads. In this phase the results from the part hash-tables are further aggregated. The buffered tables are merged using range

Table 24.1 Possible result for query in Listing 24.2

Country	Citizens
China	1,347,350,000
France	65,350,000
Germany	81,844,000
Italy	59,464,000
India	1,210,193,000
United Kingdom	62,262,000
United States	314,390,000
Japan	127,570,000

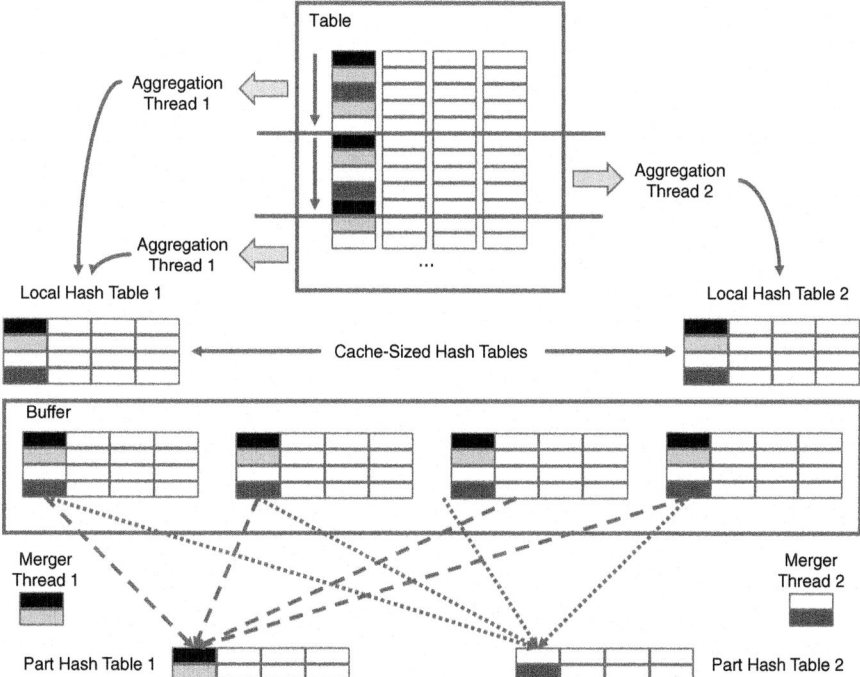

Fig. 24.1 Parallel aggregation in SanssouciDB

partitioning. Each merger thread is responsible for a certain range, indicated by a
different color in Fig. 24.1. The partitioning criterion is defined on the keys of the
local hash tables.

Considering the example in Listing 24.1, the hash values for the "gender"
attribute could be partitioned as follows: All hash keys whose binary representation
starts with "0" are assigned to one merger thread, and all keys whose binary
representation starts with "1" are assigned to another merger thread. In our example
given in Listing 24.2, the hash values for the "country" attribute are partitioned into
eight ranges. Thus, assuming that there are roughly 200 countries in the world, each

merger thread would be responsible for $\sim \frac{200}{8}$ countries if the hash function ensures a mostly uniform value distribution.

Each merger thread accesses all buffered hash tables and looks for hash values in the range that is has been assigned. Since access to the buffered hash tables is read-only, there are no contention issues due to synchronization. Similar to the hashing phase, each merger thread has a private hash table where it maintains the total number of citizens per country as it goes through all buffered hash tables from the previous phase. The result is obtained by simply concatenating the private hash tables of the merger threads.

A more detailed description of the parallel aggregation algorithm in SanssouciDB can be found in [Pla11].

24.3   Self Test Questions

1. **Aggregation - GROUP BY**
   Assume a query that returns the number of citizens of a country, e.g.:
   SELECT country, COUNT(∗)
   FROM world_population
   GROUP BY country;
   The world_population table contains the names and countries of all citizens of the world.
   The GROUP BY clause is used to express . . .

   (a) that the aggregate function shall be computed for every distinct value of country
   (b) the graphical format of the results for display
   (c) an additional filter criteria based on an aggregate function
   (d) the sort order of countries in the result set

2. **Number of Threads**
   How many threads will be used during the second (aggregation) phase of the described parallel aggregation algorithm when the table is split into 20 chunks and the GROUP BY attribute has six distinct values?

   (a) at least 10 threads
   (b) at most 6 threads
   (c) exactly 20 threads
   (d) at most 20 threads

Reference

[Pla11] H. Plattner, SanssouciDB: an in-memory database for processing enterprise workloads, in *BTW*, Kaiserslautern, ed. by T. Härder, W. Lehner, B. Mitschang, H. Schöning, H. Schwarz. LNI, vol. 180 (GI, Bonn, 2011), pp. 2–21

Part IV
Advanced Database Storage Techniques

Chapter 25
Differential Buffer

The database architecture discussed so far was optimized for read operations. In the previously described approach an insert of a single tuple can force a restructuring of the whole table if a new attribute value occurs and the dictionary has to be resorted. To overcome this problem, we will introduce the differential buffer in this chapter.

25.1 The Concept

The concept of the differential buffer (sometimes also called "delta buffer" or "delta store") divides the database into a main store and the differential buffer. All insert, update and delete operations are performed on the differential buffer plus the validity vectors. Thus, data modifications happen in the differential buffer only. The read-optimized main store is not touched by any data modifying operation. The overall current state of the data is the conjunction of the differential buffer and the main store, thus every read operation has to be performed on the main store and the differential buffer, too. Since the differential buffer is orders of magnitudes smaller than the main store, this has only a small impact on the read performance.

During query execution, a query is logically split into a query on the compressed main partition and the differential buffer. After the results of both subqueries are retrieved, the intermediate representations must be combined to build the full valid result representing the current state of the data (Fig. 25.1).

25.2 The Implementation

In the differential buffer, we keep the concept of a column-oriented layout and the use of dictionaries. However, to improve write performance, the dictionaries are not sorted but the values are stored in insertion order. Thus, resorting of the differential buffer will not occur. To speed up accesses to values in the unsorted

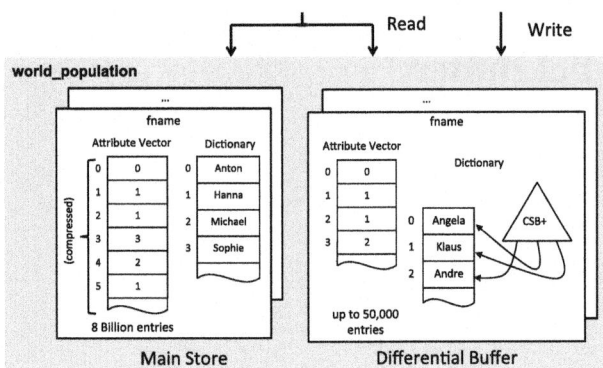

Fig. 25.1 The differential buffer concept

dictionary, we use CSB+ trees [RR00]. The CSB+ tree is mainly used as an index to look up valueIDs for a given value. While this overall approach is optimized for write performance the biggest drawback of this data storage format is that we cannot execute the queries in the exact same way as we do it in the main partition. One example are range queries. Based on the explicit order of the valueIDs in the dictionary of the compressed main partition, we can guarantee that we can represent a range of values by the valueIDs of its boundaries. Due to the insertion order of the tuples in the dictionary of the differential buffer, we have to explicitly identify each qualifying row by looking up the value in the dictionary. This is not of importance for simple single point queries that only access one tuple, but might cause a problem for the mentioned range queries [HBK+11].

The general implementation of the differential buffer is thereby as follows. First, we keep a list of all occurring values and a CSB+ tree to allow for logarithmic search in all unique values. While the unique values are not stored in a specific order as it is done in the compressed main partition, the CSB+ tree allows to define an ordering criterion on an attribute to perform fast searches on that attribute. The disadvantage of this approach is the space overhead of the tree structure. While the list of values is stored separately in simple insertion order, the tree requires additional space. The space overhead can be approximated by a factor of 2, depending on the actual fill level of the leaf nodes inside the tree.

Since read performance is critical to our mixed workload enterprise applications, we need to make sure that the differential buffer size is always kept as small as possible. Because the cost of maintaining an CSB+ tree increase with the number of distinct values in the differential buffer, we use an online reorganization process that merges the changes that are stored in the differential buffer with the compressed main partition to build a new compressed main partition. The detailed description of the merge process will follow in Chap. 27.

recId	fname	lname	gender	country	city	birthday	valid
0	Martin	Albrecht	m	GER	Berlin	08-05-1955	1
1	Michael	Berg	m	GER	Berlin	03-05-1970	0
2	Hanna	Schulze	f	GER	Hamburg	04-04-1968	1
3	Anton	Meyer	m	AUT	Innsbruck	10-20-1992	1
4	Ulrike	Schulze	f	GER	Potsdam	09-03-1977	1
5	Sophie	Schulze	f	GER	Rostock	06-20-2012	1
...	...	...	...	...	...	...	
8×10^9	Zacharias	Perdopolus	m	GRE	Athen	03-12-1979	1
0	Michael	Berg	m	GER	Potsdam	03-05-1970	1

Main Store

Differential
Buffer

Fig. 25.2 Michael Berg moves from Berlin to Potsdam

25.3 Tuple Lifetime

Because the compressed main partition of a table cannot be modified, we need a new way to identify tuple lifetime for the records stored there. The first idea that comes to mind when updating a record in the compressed main partition is that we will now perform an additional insert with the changed as well as unchanged values in the differential buffer and do nothing in the main partition. The problem that arises with this implementation is that we then can no longer distinguish between the result of the compressed main partition and the differential buffer. This is especially important if there are multiple modifications for a single record. To overcome this limitation we need to add a special system bit vector to the table that manages the validity of a tuple in the compressed main partition and the differential buffer. For each record, this validity vector stores a single bit that indicates if the record at this position is valid or not. To ensure fast read and write access to this vector, it stays uncompressed.

During query execution, the lookup of the valid tuple is handled as follows: First, the query is executed normally as it would be without the validity vector in parallel on the main partition and the differential buffer. Afterwards, when both results are available, the result positions are verified with the validity vector to remove all positions from the intermediate result that are not valid. This approach is illustrated in Fig. 25.2. In this example Michael Berg moves from Berlin to Potsdam. Since we cannot modify the main structure directly we have to execute two operations. First, we invalidate the old tuple in the main partition by clearing the valid bit, and second, we insert the complete tuple with the new location in the differential buffer so we have all information about Michael Berg available again. Both operations have to be able to be executed atomically as one single operation so that no information will be lost at any time.

The drawback of this approach is that during query execution a small additional overhead is added. However, the benefits greatly outweigh the disadvantages,

especially because using specialized SIMD instructions enables us to check multiple
positions at once for validity.

25.4   Self Test Questions

1. **The Differential Buffer**
   What is the differential buffer?

   (a) A buffer where different results for one and the same query are stored for later
       usage
   (b) A dedicated storage area in the database where inserts, updates and deletes
       are buffered
   (c) A buffer where exceptions and error messages are stored
   (d) A buffer where queries are stored until there is an idle CPU available for
       processing

2. **Performance of the Differential Buffer**
   Why might the performance of read queries decrease, if a differential buffer is
   used?

   (a) Because only one query at a time can be answered by using the differential
       buffer
   (b) Because the CPU cannot perform the query before the differential buffer is
       full
   (c) Because read queries have to query both the main store and the write-
       optimized differential buffer
   (d) Because inserts collected in the differential buffer have to be merged into the
       main store every time a read query comes in

3. **Querying the Differential Buffer**
   If we use a differential buffer, we have the problem that several tuples belonging
   to one real world entry might be present in the main store as well as in the
   differential buffer. How did we solve this problem?

   (a) This statement is completely wrong because multiple tuples for one real world
       entry never exist
   (b) We introduced a validity bit
   (c) We use a specialized garbage collector that just keeps the most recent entry
   (d) All attributes of every doubled occurrence are set to NULL in the compressed
       main store

References

[HBK+11]   F. Hübner, J.-H. Böse, J. Krüger, C. Tosun, A. Zeier, H. Plattner, A cost-aware
           strategy for merging differential stores in column-oriented in-memory DBMS, in
           BIRTE, pp. 38–52, 2011

[RR00]   J. Rao, K.A. Ross, Making B+-trees cache conscious in main memory, in *Proceedings
           of the 2000 ACM SIGMOD international conference on Management of data*,
           SIGMOD '00 (ACM, New York, 2000), pp. 475–486

Chapter 26
Insert-Only

26.1 Definition of the Insert-Only Approach

Data stored in database tables changes over time and these changes should be traceable, as enterprises need to access their historical data. Additionally, the possibility to access any data that has been stored in the database in the past and keeping historical data is mandatory e.g. for financial audits.

In this chapter, we discuss the insert-only approach. Using this approach, applications do not perform updates and deletions on the existing, physically stored data tuples, but create new tuples and manage tuple validity instead. A small glimpse of how this can be done was already provided with the introduction of the validity column in Chap. 25. By using an insert-only approach, all data changes are recorded in the same logical database table. We abstract from the introduced separation between the main store and the differential buffer here. In other words, the insert-only approach can be formulated as follows: outdated data is not overwritten or deleted, but invalidated. Invalidation can be done by means of additional attributes which indicate the currently valid revision of a tuple. This makes accessing the previous revisions of data very simple: just using the key of the tuple and the revision attribute results in the retrieval of the requested tuple in its desired version. By that, the kind of traceability that is legally required for financial applications in many countries is already provided. In addition, there are some business related benefits as well as some technical reasons for insert-only, such as:

- So called time travel queries are easily possible. Time travel queries allow users to see the data like it was at certain point in the past. Simple access to historical data enables the company management to efficiently analyze the development of the enterprise and, thus, make well-grounded strategic decisions.
- Insert-only can simplify the implementation of parallelization mechanisms, e.g. multiversion concurrency control (will be discussed below).
- In the context of in-memory, column-oriented databases that use dictionary encoding for storing the data, an insert-only approach does not require to clean the dictionaries on update operations, as the previous values remain present.

H. Plattner, *A Course in In-Memory Data Management*,
DOI 10.1007/978-3-642-55270-0_26, © Springer-Verlag Berlin Heidelberg 2014

Table 26.1 Initial state of example table using point representation

id	fname	lname	gender	country	city	birthday	valid from
0	Martin	Albrecht	m	Germany	Berlin	08-05-1955	10-11-2011
1	Michael	Berg	m	Germany	Berlin	03-05-1970	10-11-2011
2	Hanna	Schulze	f	Germany	Hamburg	04-04-1968	10-11-2011

Table 26.2 Example table using point representation after updating the tuple with id = 1

id	fname	lname	gender	country	city	birthday	valid from
0	Martin	Albrecht	m	Germany	Berlin	08-05-1955	10-11-2011
1	Michael	Berg	m	Germany	Berlin	03-05-1970	10-11-2011
2	Hanna	Schulze	f	Germany	Hamburg	04-04-1968	10-11-2011
1	Michael	Berg	m	Germany	Hamburg	03-05-1970	07-02-2012

So, how do we differentiate between the currently valid tuples and outdated ones? The following possible approaches have to be considered:

- Point representation: to determine the validity of tuples, a single field is used. The "valid from" field stores the insertion date of the tuple.
- Interval representation: to determine the validity of tuples, two fields are used. The "valid from" and "valid to" field contain information about the time interval in which the tuple was considered to be valid.

In the following, we provide use cases, implementation strategies, as well as advantages and disadvantages for both approaches. Our explanations are based on the "world_population" example. For the sake of comprehensibility, validity attributes are denoted using dates instead of timestamps with a precision of 1 ms, which would be used in a real implementation.

26.2 Point Representation

When using point representation, the "valid from" date is stored with every tuple in the database table and represents the point in time when the tuple was created. This obviously enables a fast writing of new tuples. On any update, only the tuple with the new values and the current "valid from" date has to be entered, the other tuples do not need to be changed. Consider the following initial state of the table (Table 26.1). Please note that this time the ids are stored explicitly and will be used to reference tuples (in reality, any primary key for the tuples will suffice, separate ids are not necessary).

Now we want to update the tuple with id 1, because Michael moves from Berlin to Hamburg. The update in the database table is done on 07-02-2012. In case of an insert-only approach and point representation, the update result will look as depicted in Table 26.2.

```
UPDATE world_population
SET city = 'Hamburg'
WHERE id = 1
```

Listing 26.1 Update request from user or application

```
INSERT INTO world_population
VALUES (1, 'Michael', 'Berg', 'm', 'Germany',
        'Hamburg', '03-05-1970', '07-02-2012')
```

Listing 26.2 Generated insert statement from update request

```
SELECT * FROM world_population
WHERE id = 1
ORDER BY validFrom DESC LIMIT 1
```

Listing 26.3 Point representation: retrieve most recent entry

The old tuple is kept and a new record for the tuple with the same key and a different "valid from" date is inserted. From a technical point of view, the primary key is a compound key, composed of the original primary key and the "valid from" field. To perform the update, the user or the client application issues the following SQL statement presented in Listing 26.1.

The update statement is semantically similar to the insert operation shown in Listing 26.2, if the overwriting behavior of the update is discarded, which is the case when using an insert-only approach. All attributes that are not directly specified in the update statement are retrieved from the most recent entry of the tuple.

On the downside, a point representation approach can be less efficient for read operations if the user only needs the most recent tuples. Every time when searching for the most recent tuple, the other tuples of that entry (i.e., ones having the same id) have to be fetched, as well, and all tuples need to be sorted by their "valid from" timestamp. Hence, in order to retrieve the most recent record with id = 1, the query presented in Listing 26.3 is necessary.

Due to the mentioned properties, the point representation approach is efficient for OLTP dominated workloads where write operations are required more often than read operations.

26.3   Interval Representation

When using interval representation, both "valid from" and "valid to" dates are stored with every tuple in the database table. The fields contain the creation date of a tuple and the point in time when it was invalidated.

Table 26.3 Initial state of example table using interval representation

id	fname	lname	gender	country	city	birthday	valid from	valid to
0	Martin	Albrecht	m	Germany	Berlin	08-05-1955	10-11-2011	
1	Michael	Berg	m	Germany	Berlin	03-05-1970	10-11-2011	
2	Hanna	Schulze	f	Germany	Hamburg	04-04-1968	10-11-2011	

Table 26.4 Example table using interval representation after updating the tuple with id = 1

id	fname	lname	gender	country	city	birthday	valid from	valid to
0	Martin	Albrecht	m	Germany	Berlin	08-05-1955	10-11-2011	
1	Michael	Berg	m	Germany	Berlin	03-05-1970	10-11-2011	07-02-2012
2	Hanna	Schulze	f	Germany	Hamburg	04-04-1968	10-11-2011	
1	Michael	Berg	m	Germany	Hamburg	03-05-1970	07-02-2012	

(I)   **UPDATE** world_population **SET** validTo = '07−02−2012'
         WHERE id = 1 **AND** validTo **IS NULL**

(II)   **INSERT INTO** world_population
         VALUES (1 , 'Michael', 'Berg', 'm', 'Germany',
                   'Hamburg', '03−05−1970', '07−02−2012')

Listing 26.4 Operations for update in insert only using interval representation

Similar to the point representation, interval representation requires to store the complete tuple with the "valid from" date on attribute change. Additionally, the "valid to" date field is updated in the tuple that is substituted by a newer one. This "valid to" date is equal to the "valid from" date in the new tuple. Obviously, a write operation is more complex in this case. Consider the same table as in the point representation example. The initial state is shown in Table 26.3.

Again, we want to update the tuple with the id = 1 such that the city will be changed to 'Hamburg'. In case of interval representation, the update will create the state depicted in Table 26.4.

Not only the new tuple with the updated field values is inserted, but the old, now outdated, tuple is updated, as well (see Listing 26.4). Due to this fact, insert operations are less efficient when using interval representation.

On the other hand, the additional "valid to" field simplifies read operations in comparison to point representation. In the case of an interval representation, there is no need to fetch and sort all tuples with the given id in order to get the most recent one. As shown in Listing 26.5, only the tuples with the appropriate key and an empty "valid to" date have to be selected in order to retrieve the currently valid tuple.

The mentioned properties render the interval representation approach especially efficient for OLAP dominated workloads, where read operations are required more often than write operations.

```
SELECT *
FROM world_population
WHERE id = 1 AND validTo IS NULL
```

Listing 26.5 Interval representation: Retrieve most recent entry

Table 26.5 Example table using interval representation to show concurrent updates

emplId	salary	valid from	valid to
0	10,000	10-11-2011	
1	20,000	10-11-2011	
2	15,000	10-11-2011	

26.4   Concurrency Control: Snapshot Isolation

Taking the multi-core architecture of modern CPUs and the possibility to parallelize queries into account, different ways of parallelization and concurrency control have to be investigated. As mentioned above, an insert-only approach not only helps to match compliance requirements (e.g. financial audits), but also simplifies technical aspects of an in-memory column store. Let us discover in more detail how the insert-only approach can help to simplify snapshot isolation.

The following approaches for concurrency control are commonly used:

- Locking: in this case, a transaction that locks resources works with them exclusively. An operation can be started only if all required resources are available.
- Optimistic concurrency control: in this case, data for an operation is stored isolated, in a so called virtual snapshot.

In the second approach, all manipulations are performed on the so-called virtual snapshot, i.e., the data that is valid at the time the transaction started. So, when using this variant of multiversion concurrency control, transactions that need to update data actually insert new versions into the database, but concurrent transactions will still see a consistent state based on previous versions, because they all work on their own "virtual copy" of the data. Obviously, that can cause conflicts.

The following example outlines how the insert-only approach can simplify multiversion concurrency control. Consider the following simple table that stores the salary of employees and uses interval representation to implement the insert-only approach. Table 26.5 shows its state at time T_1.

Now, two concurrent transactions operate on the data stored in this table. Transaction 1 starts at time T_1. It reads the salary value for the employee with "emplId" = 0. At T_1 this value is "10,000". At time T_3, the transaction updates the salary value for the respective employ to "12,000", which results in the state depicted in Table 26.6. A new tuple is being created with a "valid from" value that equals the timestamp (or the date in this simplified example) of T_3, "07-07-2012". The previously valid tuple is invalidated by setting the "valid to" attribute, respectively.

Table 26.6 Example table using interval representation to show concurrent updates after first update

emplId	salary	valid from	valid to
0	10,000	10-11-2011	07-07-2012
1	20,000	10-11-2011	
2	15,000	10-11-2011	
0	12,000	07-07-2012	

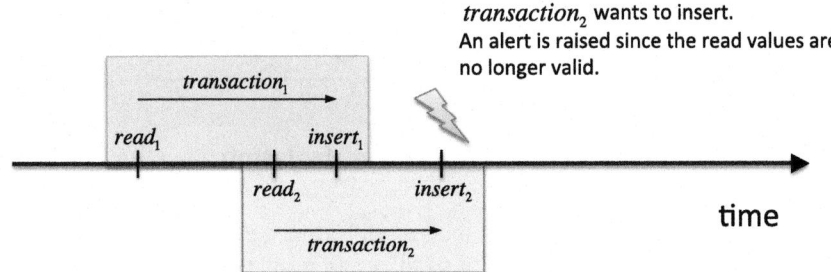

Fig. 26.1 Snapshot isolation

The second concurrent transaction starts at time T_2, with T_2 happening in between T_1 and T_3. It also operates on the record where "EmplId" $= 0$. At time T_2, the second transaction obviously cannot access the updates done by the first transaction at time T_3, since they did not happen, yet. Hence, it reads the original value of "10,000" for the salary. If the second transaction now tries to update the record at time T_4, which happens after T_3, the concurrency control can determine whether the last read value originates from the currently valid tuple or has been modified in the meantime, and act accordingly. The described sequence of events is visually summarized again in Fig. 26.1.

26.5 Insert-Only: Advantages and Challenges

As described above, we never delete data from a table. This raises one question: "How does the insert-only approach influence memory consumption?". For each row update, an additional tuple including the additional timestamp (or timestamps when using interval representation) have to be inserted into the database and the stored data volume will increase accordingly. But how does this affect memory consumption in real world scenarios? To answer this question, we need to consider which types of updates are usually performed in business applications:

- Aggregate updates
- Status updates
- Value updates

Taking the advantages of an in-memory column-based database into account, most aggregates can be efficiently calculated on the fly. Others can be processed and accelerated by the aggregate cache (cf. Aggregate Cache Chapter). Therefore, materialized aggregates do not need to be maintained or updated. Concerning the remaining update types, a study was conducted at HPI [KKG$^+$11]. It showed that a typical SAP financials application is not update-intensive: only an average of about 5 % of all operations are updates (see Chap. 3). This already reduces the impact of the insert-only approach on memory consumption, as only a small amount of data is updated, at all. Adding the fact that most of these updates are status updates, another smart implementation strategy helps to reduce the impact on memory consumption even further. Most status fields only consist of one bit of information, which means that if a status is being set (e.g., the "delivered" field of an order), the respective attribute can directly be replaced with the timestamp of the change. When doing the status update this way, all information is encapsulated in one field and the update can be done in place, so no additional record has to be written. These characteristics and improvements lead to the result that the memory consumption only increases moderately when insert-only is being used.

26.6 Self Test Questions

1. Statements Concerning Insert-Only
Considering an insert-only approach, which of the following statements is true?

(a) Using an insert-only approach, invalidated tuples can no longer be used for time travel queries
(b) Old data items are deleted as they are not necessary any longer
(c) Data is not deleted, but invalidated instead
(d) Historical data has to be stored in a separate database to reduce the overall database size

2. Benefits of Historical Data
Consider keeping deprecated or invalidated (i.e. historical) tuples in the database. Which of the following statements is wrong?

(a) It is legally required in many countries to store historical data
(b) Analyses of historical data can be used to improve the scan performance.
(c) Historical data can provide snapshots of the database at certain points in time
(d) Historic data can be used to analyze the development of the company

3. Accesses for Point Representation
Considering point representation and a table with one tuple, that was invalidated five times, how many tuples have to be checked to find the most recent tuple?

(a) Two, the most recent one and the one before that
(b) Five

(c) Six

(d) Only one, that is, the first which was inserted

4. Statement concerning Insert-Only

Which of the following statements concerning insert-only is true?

(a) Interval representation allows more efficient write operations than point representation

(b) In interval representation, four operations have to be executed to invalidate a tuple

(c) Point representation allows faster read operations than interval representation due to its lower impact on tuple size

(d) Point representation allows more efficient write operations than interval representation

Reference

[KKG+11] J. Krueger, C. Kim, M. Grund, N. Satish, D. Schwalb, J. Chhugani, H. Plattner, P. Dubey, A. Zeier, Fast updates on read-optimized databases using multi-core CPUs. Proc. VLDB **5**, 61–72 (2011)

Chapter 27
The Merge Process

Using a differential buffer as an additional data structure to improve write performance requires to periodically combine this data with the compressed main partition. This process is called "merge".

The reasons for merging are two fold. On the one hand, merging the data of the differential buffer into the compressed main partition decreases the memory consumption due to better compression. On the other hand, merging both data structures improves the read performance due to the sorting of the dictionary of the read-optimized main store. In an enterprise context, the requirements of the merge process are that it

- can be executed asynchronously,
- has as little impact as possible on all other operations, and
- does not block any OLTP or OLAP transactions.

To achieve this goal, the merge process creates a new empty differential buffer and a copy of the main store prior to the actual merging phase to avoid locking during the merge. Incoming data modifications are passed to the new differential buffer. Using this approach, we are able to reduce the time that we have to explicitly lock the compressed main store and the differential buffer. Using the copy of the main store and the new differential buffer, the merged table is locked for the short time period when the pointer to the main store of a table is set to the new main store after the merge. In this online merge concept, the table is available for reads and the second differential buffer for write operations during the merge process.

During the merge, the process requires additional system resources (CPU and main memory) that have to be considered during system sizing and scheduling.

For the differential buffer concept, it is important to mention that all update, insert, and delete operations are captured as technical inserts into the differential buffer while a dedicated valid tuple vector per table and store ensures consistency.

H. Plattner, *A Course in In-Memory Data Management*,
DOI 10.1007/978-3-642-55270-0_27, © Springer-Verlag Berlin Heidelberg 2014

When using a differential buffer, the update, insert, and delete performance of the database is limited by two factors:

- the insert rate into the write-optimized data structure and
- the performance with which the system can merge the accumulated modifications into the read-optimized main store.

By introducing a differential buffer, the read performance is decreased depending on the number of tuples in the differential buffer. This impacts especially Join operations since they heavily rely on sorted dictionaries. This means that intermediate values from the differential buffer cannot be directly compared to those from the compressed main partition. This can force the execution engine to switch to an execution based on early materialization which can have a severe performance impact (see Chap. 16). Consequently, the merge process has to be executed if the performance impact becomes too large. It is triggered by one of the following events:

- The number of tuples in the differential buffer for a table exceeds a defined threshold.
- The memory consumption of the differential buffer exceeds a specified limit.
- The differential buffer log for a columnar table exceeds the defined limit.
- The merge process is triggered explicitly by a SQL command.

27.1   The Asynchronous Online Merge

To enable the execution of queries during a running merge operation (online applicability), we introduce the concept of an asynchronous merge. The overall requirement for this process is that it will not block any concurrent modifying transactions. Figure 27.1 illustrates this concept.

By introducing a second differential buffer, data changes on the table can still be applied, even during the merge process. Consequently, read operations have to access both differential buffers to query the current of tuples state during the merge. To maintain consistency, the merge process requires a lock at the beginning and at the end of the process of switching the stores and applying necessary data modifications—such as valid tuple modifications—which have occurred during the merge process. Open transactions are not affected by the merge, since their changes are copied from the old into the new differential buffer and can be processed in parallel to the merge process. To finish the merge operation, the old main store is replaced with the new one. Within this last step of the merge process, a snapshot of the new main store is persisted, which defines a new starting point for log replay in case of failures (see Chap. 29).

The merge process consists of three phases: (1) prepare merge, (2) attribute merge, and (3) commit merge. Phase (2) is hereby carried out for each attribute of the table.

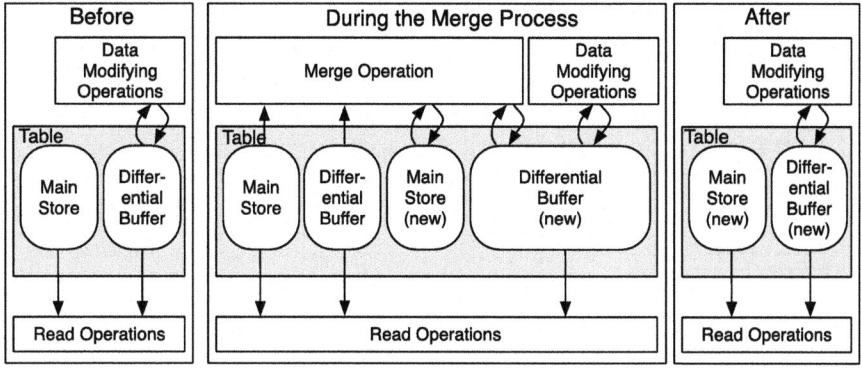

Fig. 27.1 The concept of the online merge

27.1.1 Prepare Merge Phase

First, theprepare merge phase creates a new empty differential buffer for all new inserts, updates, and deletes that occur during the merge process. Then, the differential buffer as well as the main store are locked. Additionally, the current validity vectors of the old differential buffer and the main store are copied because these may be changed by concurrent updates or deletes applied during the merge, which may affect tuples involved in this process.

27.1.2 Attribute Merge Phase

The attribute merge phase as outlined in Fig. 27.2 consists of two steps. In the first step, the differential buffer and main store dictionaries are combined into one sorted result dictionary. In addition, a value mapping is created as an auxiliary mapping structure to map the positions from the old dictionary to the new dictionary for the differential buffer and the main store. These auxiliary structures are actually not necessary for the algorithm, but avoid expensive lookups in the old dictionaries and improve cache utilization.

The input dictionaries to be consolidated are the main store's dictionary and the sorted differential buffer dictionary build from the differential buffer's unsorted dictionary using the additional CSB+ tree. Having the sorted dictionaries, both are merged and form the resulting dictionary containing the main store's and differential buffer's distinct values.

In the second step, the values from the two attribute vectors are copied into a new combined attribute vector. Therefor the auxiliary structure is used. To ensure that the sizing of the new attribute vector is correct, we calculate the required value width based on the size of the new dictionary.

An exemplary run of the attribute merge phase is shown in Sect. 27.2.

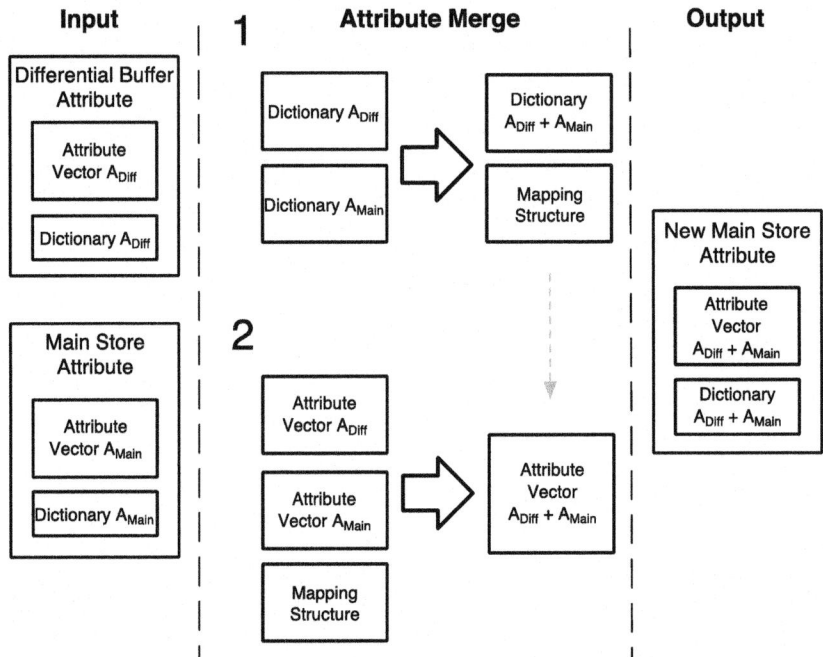

Fig. 27.2 The attribute merge

27.1.3   Commit Merge Phase

The commit merge phase starts by acquiring a write lock of the table. This ensures that all running queries are finished before the switch to the new main store including the updated valueIDs takes place. Then, the valid tuple vector that was copied in the prepare phase is applied to the actual vector to mark invalidated tuples. As the last step, the new main store replaces the original differential buffer as well as the old main store and the memory allocations of both are freed.

The result of the merge process for a simple example is shown in Fig. 27.3. The new attribute vectors hold all tuples of the original main store, as well as the ones from the differential buffer. Note that the new dictionaries include all values from the main store and the differential buffer and they are sorted to allow for binary search and efficient range queries.

27.2   Exemplary Attribute Merge of a Column

The Attribute Merge described in Sect. 27.1 will be explained with a simplified example, showing the merge of a single column. Please note that this does include the optimizations of the single column merge concept described in Sect. 27.3.2. The

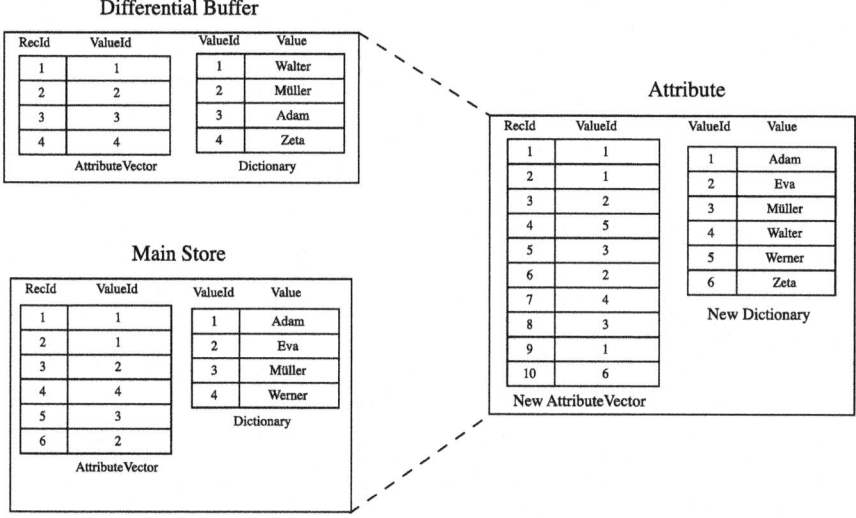

Fig. 27.3 The attribute vector and dictionary of the main store after the merge process

overall process consists of two distinct steps, which are shown in Figs. 27.4 and 27.5.

The first step takes the dictionary of the main store's attribute vector (here denoted as D_m) of the main store column (AV_m) and its counterpart D_d. Having a CSB+ tree on top of the unsorted differential buffer dictionary (see Fig. 25.1), we can iterate over D_d in a sorted order and therefore pictured it in a sorted order here. To merge both dictionaries into a combined sorted dictionary D_c, first the pointer on both dictionaries are set to the first element. The values of the pointers of both dictionaries are then compared in each iteration, the smaller value is added to the result, and the corresponding pointer is incremented. In the event that both values are equal, the value is added once and both pointers are incremented. To be able to later update the attribute vector to the newly created dictionary, every time a value is added to the combined dictionary D_c, the mapping information from the corresponding old dictionary to the combined one is added to the corresponding auxiliary structures AUX_m and AUX_d. If one of the two pointers reaches the end of its input dictionary, the remaining items of the other dictionary are copied directly to the end of the combined dictionary, since both dictionaries are sorted.

That means, that at the beginning of this step, the pointer on D_m marks "anna", and the pointer on AV_d marks "bernd". As "anna" is lexically located before "bernd", it is added to D_c first and the pointer on D_m is advanced to "charlie". Afterwards, "bernd" is added to D_c and the pointer on the sorted list of values of the CSB+ tree is advanced to "frank". As can be seen, entries (like "anna", arrow M1) only present in the old main store dictionary, entries only present in the CSB+ tree of the differential buffer ("bernd", arrow D1) and entries present in both structures ("frank", arrow B1) get transferred to the new combined dictionary D_c.

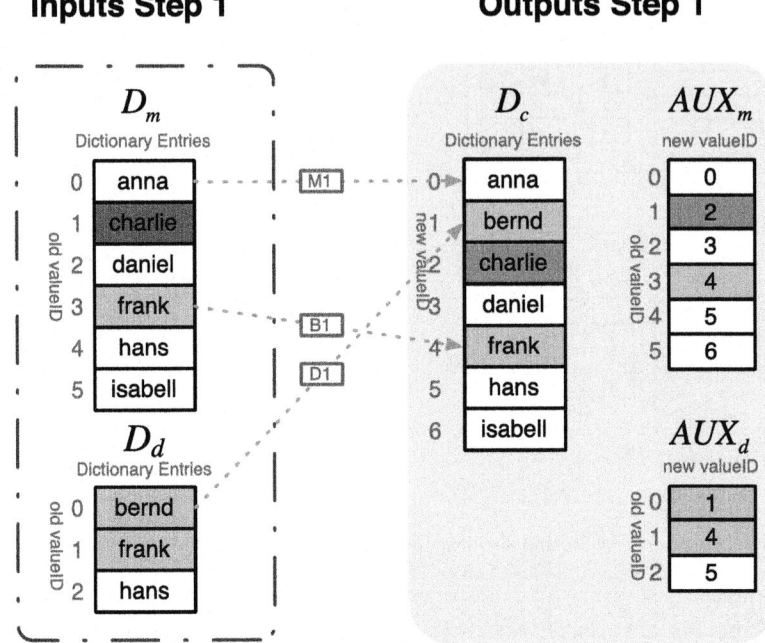

Fig. 27.4 First step of the merge process for a single column: dictionary merge (adapted from [FSKP12])

While constructing the combined dictionary, the auxiliary structures AUX_m and AUX_d are filled with the resulting new valueIDs which will most probably differ from the old valueIDs. This is necessary for both structures, not just the differential buffer: for the main store, the new valueIDs might be increased in comparison to the old ones due to the addition of new entries from the differential buffer.

The second step builds up the combined attribute vector AV_c using the attribute vector AV_m of the old main store, the leafs of the CSB+ tree (AV_d) which represent the attribute vector of the differential buffer in a sorted order and the just created auxiliary structures AUX_m and AUX_d. The valueIDs of the outdated attribute vectors are sequentially scanned. Each value is updated with the help of the appropriate auxiliary structure and then just added to AV_c. The arrows labelled M2 show an example for the entry "charlie" from the main store, which was represented by the valueID 1, but is now represented by the valueID 2 due to the addition of the value "bernd" to the combined dictionary. The arrows labelled D2 show an analog example for the value "bernd" which comes from the differential buffer. All in all, the resulting combined attribute vector is the concatenation of the existing attribute vectors with updated valueIDs.

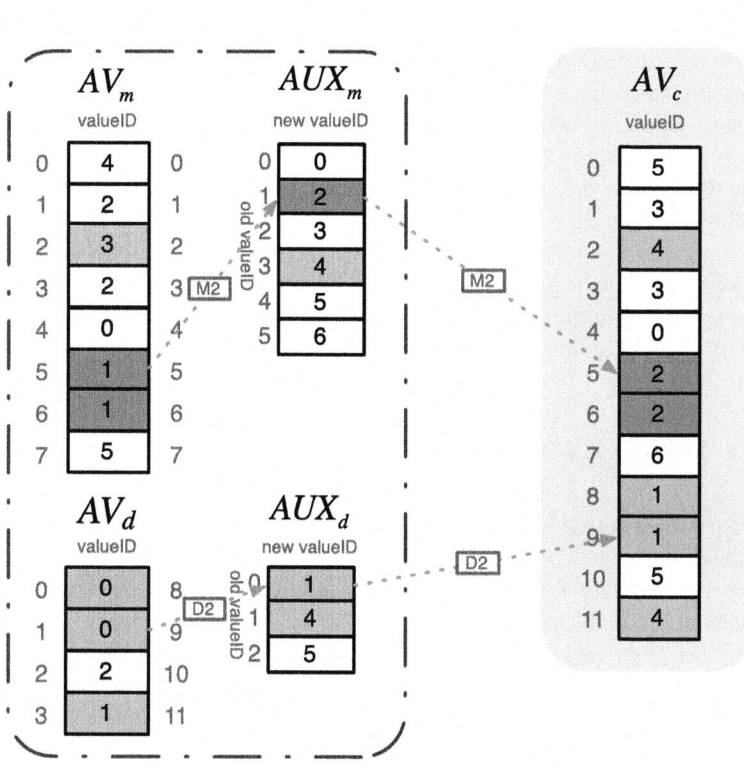

Fig. 27.5 Second step of the merge process for a single column: attribute vector rewrite (adapted from [FSKP12])

27.3 Merge Optimizations

In addition to the described asynchronous online merge, this section presents further optimizations.

27.3.1 Using the Main Store's Dictionary

The first optimization is the usage of the main store's dictionary in the differential buffer. One of the major advantages of writing to a differential buffer is that adding new elements can be done without the eventual penalty of re-sorting the dictionary or re-encoding the attribute vector. A disadvantage is that an additional dictionary is created which has to be incorporated into the main store's dictionary during the merge. But in several cases, using the main store's dictionary in the differential buffer as well can improve the merge performance. This is the case when the main store's dictionary is already saturated. Typical examples are columns storing years, gender, country- or postal codes. These columns are saturated early and new elements are rather rare.

In these cases, all elements in the differential buffer already use the final dictionary positions of the main store. This can severely reduce the memory consumption of the differential buffer. Consequently, the merge of an attribute that reuses the main dictionary is a simple concatenation of the differential buffer tuples to the main store's attribute vector. In cases where the probability of data modifications introducing new values to the dictionary is high, using the main store's dictionary in the differential buffer is rather expensive. This is the case for attributes such as timestamps, entity identifier (IDs), or similar.

27.3.2 Single Column Merge

Another possible optimization concerns memory consumption. During the merge phase, the complete new main store is kept in main memory. At the point of highest memory consumption, more than twice the size of the original main store plus the size of the differential buffer is required to be stored in main memory to execute the proposed merge process. Tables in enterprise applications often consist of millions of tuples while having hundreds of attributes. As a consequence, requiring full table copies can lead to a huge overhead since at least twice the size of the largest table has to be available in memory to allow the merge process to run. For example, the financial accounting table of a large consumer products company contains about 250 million line items with 300 attributes. The uncompressed size with variable length fields of the table is about 250 GB and can be compressed with bit compressed dictionary encoding to 20 GB [KGZP10]. However, to run the merge process, at least 40 GB of main memory are necessary.

To avoid storing a complete table in memory twice, Krueger et al. present the so-called Single Column Merge that merges a table column-wise [KGW+11]. Consequently, not the whole table needs to be kept in memory twice, but only a single column. Thus, if all columns are merged sequentially, the required amount of memory is reduced to the size of the differential buffer and the compressed table, plus the size of the largest resulting column. A drawback of this approach is, that querying as well as transaction management on a partially merged table becomes more complex.

27.3.3 Unified Table Concept

To further improve the transactional capabilities of column stores, in [SFL+12] Sikka et al. present a modified differential buffer concept, called Unified Table Con-

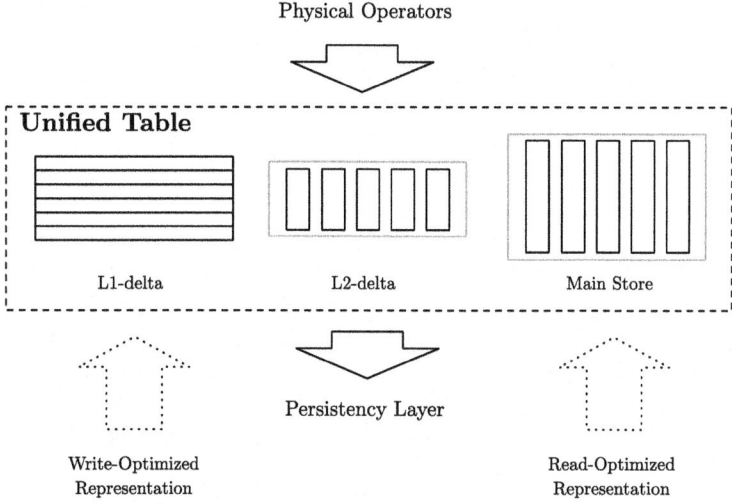

Fig. 27.6 Unified table concept (adapted from [SFL⁺12])

cept. Hereby, an additional data structure in form of an in-memory row store—called L1-delta—is used (see Fig. 27.6), while the L2-delta (i.e. the differential buffer in SanssouciDB) and the main store have similar structures as in SanssouciDB.

In this concept, each data modification is first written to the L1-delta. This structure stores approximately 10,000–100,000 rows. The L1-delta is merged with the L2-delta at regular intervals or when a certain row-limit is reached. The L2-delta is suited to store up to ten million rows and is merged with the main store. Additionally, this approach introduces bulk-loading improvements.

27.4 Self Test Questions

1. What is the Merge?
  The merge process . . .

  (a) combines the main store and the differential buffer to increase the parallelism
  (b) merges the columns of a table into a row-oriented format
  (c) optimizes the write-performance
  (d) incorporates the data of the write-optimized differential buffer into the read-optimized main store

2. When to Merge?
  When is the merge process triggered?

  (a) Before each SELECT operation
  (b) When the space on disk runs low and the main store needs to be further compressed

(c) After each INSERT operation
(d) When the number of tuples within the differential buffer exceeds a specified threshold

References

[FSKP12]   M. Faust, D. Schwalb, J. Krueger, H. Plattner, Fast lookups for in-memory column stores: group-key indices, lookup and maintenance, in *ADMS '12: Proceedings of the 3rd International Workshop on Accelerating Data Management Systems Using Modern Processor and Storage Architectures at VLDB'12*, 2012
[KGZP10]   J. Krueger, M. Grund, A. Zeier, H. Plattner, Enterprise application-specific data management, in *EDOC*, pp. 131–140, 2010
[KGW+11]   J. Krueger, M. Grund, J. Wust, A. Zeier, H. Plattner, Merging differential updates in in-memory column store, in *DBKDA*, 2011
[SFL+12]   V. Sikka, F. Färber, W. Lehner, S.K. Cha, T. Peh, C. Bornhövd, Efficient transaction processing in SAP HANA database: the end of a column store myth, in *SIGMOD Conference*, pp. 731–742, 2012

Chapter 28
Aggregate Cache

OLTP and OLAP systems (cf. Sect. 3.2) employ different approaches to deal with aggregate queries. While OLAP systems make extensive use of materialized views [SDJL96, ZGMH+95], we see that the handling of aggregates in OLTP systems is often done within the application by maintaining predefined summary tables. This leads to an increased application complexity with risks for violating data consistency and to a limited throughput of insert and update queries as the related summary tables must be updated in the same transaction [JMS95, Pla09].

With the ongoing trend of columnar IMDBs, this artificial separation of OLTP and OLAP is not necessary anymore as they are capable of handling mixed workloads, with transactional and analytical queries in a single system [Pla09]. Despite the aggregation capabilities of columnar IMDBs, access to tuples of a materialized aggregate - which we define as a materialized view whose creation query contains aggregate functions [SS77] - is always faster than aggregating on the fly. However, the overhead of materialized view maintenance to ensure consistency for modified data has to be considered and involves several challenges [GM95].

While existing materialized view maintenance strategies are applicable in columnar IMDBs [MBH+13], their main-delta architecture is well-suited for a novel strategy of caching aggregate queries and applying incremental view maintenance techniques [MP13]. In columnar IMDBs, the storage can be separated into a highly compressed, read-optimized *main* storage and a write-optimized *differential* buffer. New records are inserted to the differential buffer and periodically *merged* to the main storage [KKG+11]. With this architecture, materialized aggregates do not have to be invalidated when new records are inserted to the differential buffer, because the materialized aggregates are only defined on records from the main storage. The incremental materialized view maintenance of aggregates is decoupled from query processing and takes place during the delta merge process.

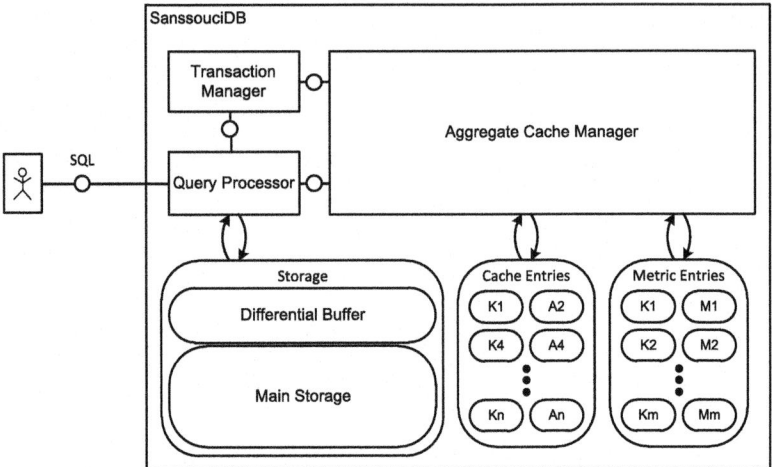

Fig. 28.1 Aggregates query caching architecture

28.1   Architecture

The architecture of the aggregate cache is illustrated in Fig. 28.1. Its main component is the *aggregate cache manager*. It maintains two maps for the caching of aggregate queries. The first one stores *cache entries*, the second one keeps *metric entries*. Each cache entry has a cache key that maps to a cached aggregate. A cache key is a unique identifier that allows unambiguous mapping between an aggregate query and a cached aggregate. A cached aggregate contains the query result calculated only on the main storage. A metric entry has the same key as the cached aggregate to which it belongs. The key points to meta data about each cached aggregate such as the calculation time, the access history, and the aggregate's size. The metrics are required for cache eviction decisions and the incremental revalidation of cached aggregates.

As shown in Fig. 28.1, the aggregate cache manager interacts with the *query processor* and the *transaction manager*. In order to support the revalidation of cached aggregates, the cache manager requires information about deleted records from the transaction manager for the revalidation of cached aggregates. The cache manager closely works together with the query processor to obtain the query results from the main storage and the differential buffer. Their interaction is described in Fig. 28.2, which shows the query processing when the aggregate cache is active.

In the first step, the query processor parses the query. In case the query does not contain any aggregations, it is executed without the cache. If a supported aggregate function is present in the query, the query processor passes the query to the aggregate cache manager. It checks, whether the aggregate to which the query belongs is already cached. In case it is not cached, the cache manager triggers the

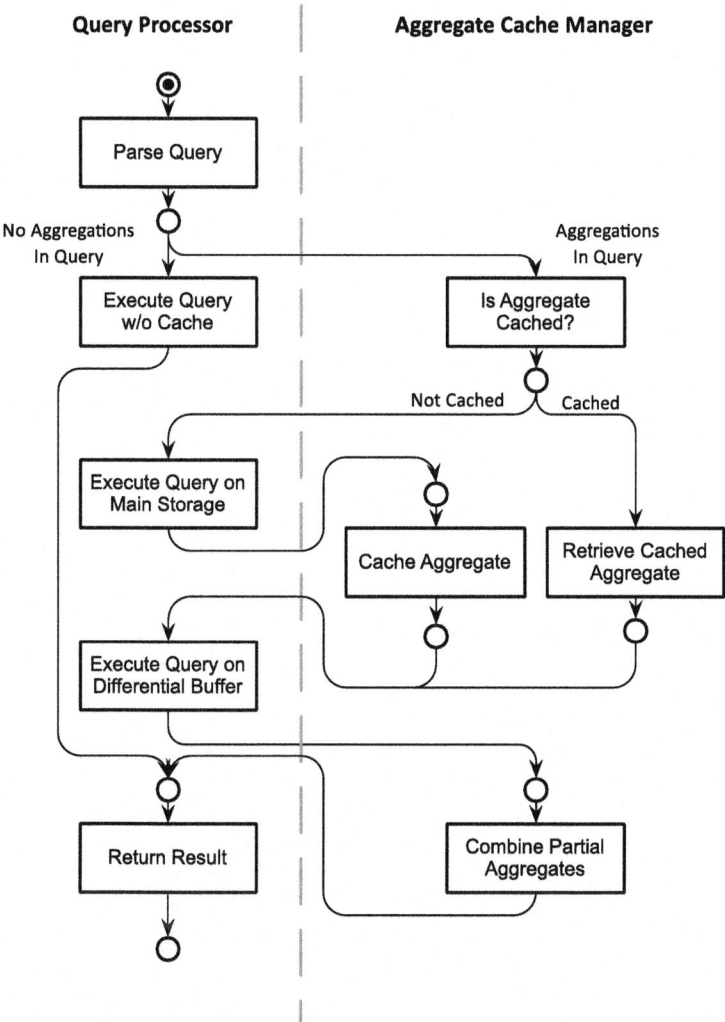

Fig. 28.2 Simplified query caching algorithm

execution of the query only on the main storage and adds the returned aggregate to
the cache entries. If the aggregate is already cached, the cache manager obtains it
from the matching cache entry. Irrespective of whether a cache hit or a cache miss
has occurred, the cache manager triggers the query execution on the differential
buffer as the next step. The cache manager combines the result from that execution
with the aggregate from the main storage in order to provide an up-to-date result
to the query processor, which returns the result as final step. The query processing

flow shown in Fig. 28.2 does not contain extensions required for the revalidation of cached aggregates or the cache management.

28.2   Cache Management

Some aggregates are accessed only once or yield hardly any performance benefit when they are cached. These aggregates only content the main memory and should not be cached. The aggregate cache manager uses a profit metric to assess the benefit of each cached aggregate and identify the unprofitable aggregates to be evicted from the cache. The profit metric considers the aggregate's size, access history, and calculation time on the main storage, as they have the highest influence on the benefit of a cached aggregate. The eviction process runs asynchronously to query processing in order to avoid blocking behavior during query processing. It evicts aggregates with the lowest assigned profit first and continues until the cache size falls below a predefined threshold.

28.3   Updates and Deletes

Cached aggregates are based on a certain database snapshot of the main storage. As long as there are no delete or update queries affecting the main storage, the cached aggregates can be reused. However, a transactional workload is characterized by theses types of queries and therefore modifies the main storage. As a result, the changes have to be propagated to the cached aggregate. The transaction manager of SanssouciDB (cf. Fig. 28.1) provides a mechanism that enables an efficient determination of changes between the creation of the cached aggregate and the time of the access. In Table 28.1, the location table is shown with the visibility information of the transaction manager. A bit vector column contains the information whether a record is visible (value 1) or not (value 0). In the example, the record with recID 2 is not visible anymore and therefore not part of the cached aggregate. All other records that are visible are part of the cached aggregate. To capture that snapshot of the database, the bit vector is copied to the cache entry. In case a record in the main storage is invalidated, the current bit vector of the transaction manager and the copied bit vector of the cache entry are compared with an XOR operation. The resulting bit vector contains all records that have been changed. With this information, it is possible to incrementally maintain the cached aggregate by incorporating the determined changes. Finally, the bit vector of the cache entry has to be updated with the current bit vector. This process is apparently more efficient compared to a full recalculation of the aggregate query which includes a full table scan.

Table 28.1 Visibility information of the location table

recID	City	State	Country	Area	Bit vector
1	Berlin	Berlin	Germany	851	1
~~2~~	~~Potsdam~~	~~Brandenburg~~	~~Germany~~	~~187~~	0
...	...	...	...	...	...
812	Kassel	Hessen	Germany	107	1

28.4 Joins

One challenge of the aggregate caching mechanism and the involved incremental materialized view maintenance is to handle aggregate queries that are based on joins of multiple tables. These queries require a union of joining all permutations of delta and main partitions of the involved tables, excluding the already cached joins between the main partitions. For a query joining two tables, three extra subjoins are required, and a query joining three tables already requires seven extra subjoins. This may result in very little performance gains over not caching at all the query on the main partitions. However, after analyzing the characteristics of enterprise applications, we identified several schema design and data access patterns that can be leveraged to allow the pruning of certain subjoins. For example, if we know that matching records are always inserted within the same transaction, and both tables are merged at the same time, we can prune all subjoins between delta and main partitions because matching tuples are either in the delta or main partitions. This can optimize the overall performance of join queries using the aggregate cache by order of magnitudes.

28.5 Self Test Questions

1. Aggregate Entries

What does the aggregate cache store in the aggregate entries?

(a) Aggregate query results computed on the main storage
(b) Aggregate query results computed on the main storage and the differential buffer
(c) Any type of query result
(d) Aggregate query results computed on the differential buffer

2. Metric Entries

For what purpose is the information in the metric entries mainly used?

(a) Aggregations on the differential buffer
(b) Query plan optimization
(c) Unique identification of aggregate entries
(d) Cache eviction decisions

3. Query Types
For which query types is the aggregate cache best suited?

(a) Recurring analytical queries
(b) Transactional queries
(c) All types of queries
(d) Distinct queries with aggregate functions

References

[GM95]   A. Gupta, I.S. Mumick, Maintenance of materialized views: problems, techniques,
             and applications. IEEE Data Eng. Bull. **18**(2), 3–18 (1995)
[JMS95]   H.V. Jagadish, I.S. Mumick, A. Silberschatz, View maintenance issues for the
             chronicle data model, *PODS* (ACM, New York, 1995), pp. 113–124
[KKG⁺11]   J. Krueger, C. Kim, M. Grund, N. Satish, D. Schwalb, J. Chhugani, H. Plattner,
             P. Dubey, A. Zeier, Fast updates on read-optimized databases using multi-core
             CPUs. PVLDB (2011)
[MBH⁺13]   S. Müller, L. Butzmann, K. Howelmeyer, S. Klauck, H. Plattner, Efficient view
             maintenance for enterprise applications in columnar in-memory databases, in *14th
             IEEE International Enterprise Distributed Object Computing Conference (EDOC)*,
             pp. 249–258, 2013
[MP13]   S. Müller, H. Plattner, Aggregates caching in columnar in-memory databases, in *1st
             International Workshop on In-Memory Data Management and Analytics (IMDM)
             at VLDB 2013*, August 2013
[Pla09]   H. Plattner, A common database approach for OLTP and OLAP using an in-
             memory column database, in *SIGMOD*, pp. 1–2, 2009
[SDJL96]   D. Srivastava, S. Dar, H.V. Jagadish, A.Y. Levy, Answering queries with
             aggregation using views, in *VLDB*, pp. 318–329, 1996
[SS77]   J.M. Smith, D.C.P. Smith, Database abstractions: aggregation. Comm. ACM,
             105–133 (1977)
[ZGMH⁺95]   Y. Zhuge, H. Garcia-Molina, J. Hammer, J. Widom, Y. Zhuge, H. Garcia-Molina,
             J. Hammer, J. Widom, View maintenance in a warehousing environment. *ACM
             SIGMOD Rec.* (ACM, New York, 1995), pp. 316–327

Chapter 29
Logging

To be used in productive enterprise applications, databases need to provide durability guarantees (as part of the ACID[1] principle). To provide these guarantees, fault-tolerance and high availability have to be ensured. However, since hardware failures or power outages cannot be avoided or foreseen, measures have to be taken which allow the system to recover from failures.

The standard procedure to enable durable recovery is logging. With the help of logging and recovery protocols, databases can be brought back to the last consistent state before the failure. This is achieved by check pointing the current system and logging subsequent data modifications. The data is written into log files, which are stored on persistent memory such as hard disk drives (HDD) or solid-state drives (SSD).

Please note that these requirements are true for any database, regardless of being an in-memory database or not.

29.1 Logging Infrastructure

A key consideration when talking about logging is performance, both for writing the logs as well as for reading logs back into memory when recovering. As discussed in Sect. 4.6, the performance gap between disk and CPU is steadily increasing. Consequently, logging has to be primarily optimized with respect to minimizing I/O operations.

Figure 29.1 outlines the logging infrastructure of SanssouciDB. The logging data, which is written to disk, consist of three parts:

---

[1]ACID stands for Atomicity, Consistency, Isolation, Durability. These properties guarantee reliability for database transactions and are considered as the foundation for reliable enterprise computing.

H. Plattner, *A Course in In-Memory Data Management*,
DOI 10.1007/978-3-642-55270-0_29, © Springer-Verlag Berlin Heidelberg 2014

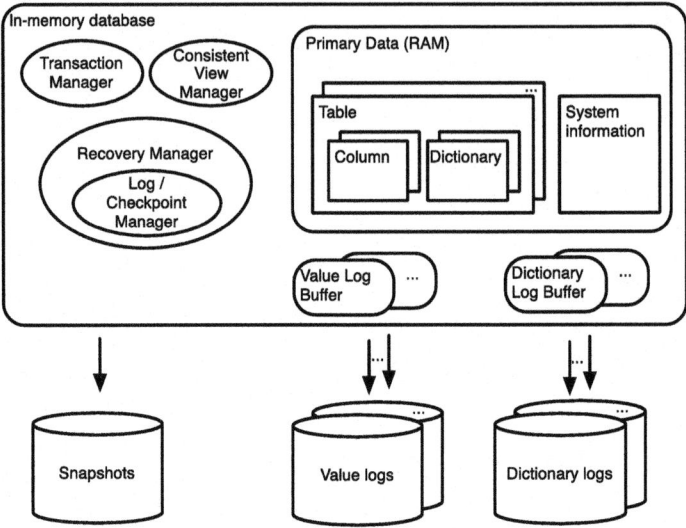

Fig. 29.1 Logging infrastructure

- Snapshot of the main store
- Value logs
- Dictionary logs

Check pointing [Bor84] is used to create a snapshot of the database at a certain point in time, at which the data is in a consistent state. According to [HR83], a database is in a consistent state "if and only if it contains the results of all committed transactions". The snapshot is a direct copy of the read-optimized main store and it is written to disk periodically. The purpose of check pointing is to speed up the recovery process, since only log entries after the snapshot have to be replayed, while the main store can be loaded from the snapshot directly. To log the data of the differential store, which is not part of the snapshot, value logs as well as dictionary logs are used to track committed changes.

SanssouciDB's logging infrastructure differs from most traditional databases. SanssouciDB adapted the infrastructure to leverage the columnar data structures and to reduce I/O performance penalties. Amongst these optimizations are:

- **Snapshot format:** at each checkpoint, a snapshot of the main store is directly written to disk in binary format. This means that an exact copy of the main store in memory is written to disk, which can later be directly restored without any transformation overhead in case of a recovery.
- **Checkpoint timing:** the ideal timing for check pointing is when the differential buffer is relatively small compared to the main store. That is right after the merge.
- **Storing meta data:** to speed up the recovery process, additional meta data is written to disk. With the help of this meta data, the required memory can be

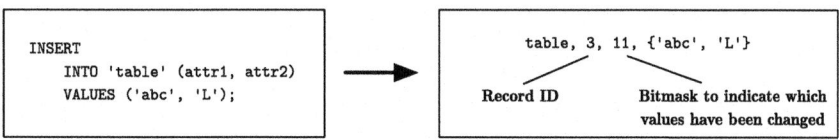

Fig. 29.2 Logical logging

allocated before loading. Thus, expensive re-allocations and data movements can be avoided. Data hereby written is, e.g., the number of tuples in the main store and the number of bits used in each dictionary.

- **Separation into value and dictionary logs:** the two major performance optimizations for logging in SanssouciDB are the reduction of the log size and the parallelization of logging. This is achieved by using dictionary-encoded logging, which is discussed in detail in the next section.

29.2 Logical vs. Dictionary-Encoded Logging

The obvious way to log data modifications is logical logging. As depicted in Fig. 29.2, logical logging simply writes the SQL statement and its parameters (recordID and attribute values) to disk.

Logical logging has two major shortcomings. First, logging and recovery cannot be parallelized since the order of the log has to be preserved during replay to recover the dictionary and the corresponding attribute vector elements. Second, logical logging writes values directly to disk and, therefore, does not leverage compression as used in SanssouciDB. Consequently, logical logging writes comparatively large data volumes to disk.

To avoid these shortcomings, SanssouciDB uses the so-called dictionary-encoded logging schema [WBR$^+$12], which separates the dictionary-encoded data and their corresponding dictionary inserts from the transactional context. This approach allows writing and recovering of the attribute vectors and dictionaries in parallel. Furthermore, dictionary-encoded logging reduces the log size due to the usage of dictionary-compression, which speeds up the recovery process significantly.

In which cases dictionary-encoded logging is advantageous over logical logging depends on the data characteristics. In enterprise applications, the same data characteristics that favor dictionary-compressed column-stores also apply to dictionary-encoded logging. Amongst these characteristics are, e.g., a low number of distinct values, which leads to fewer dictionary log entries, and the distribution of values.

Many real world value distributions can be described using the Zipf distribution. Intuitively, the Zipf distribution describes—depending on the variable *alpha*—how heavily the distribution is drawn to one value. In the case of *alpha = 0* the

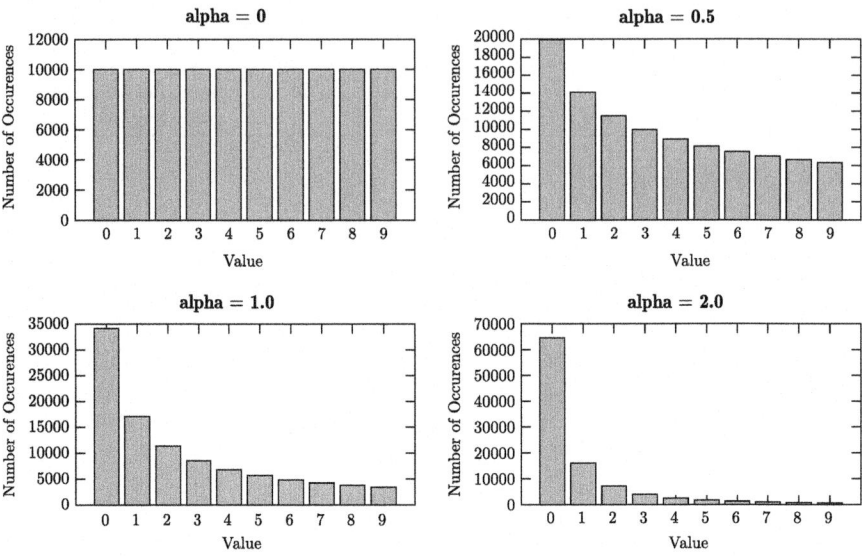

Fig. 29.3 Exemplary Zipf distributions for varying alpha values

distribution equals a uniform distribution and every value occurs equally often. As alpha increases, fewer values occur more frequently (see exemplary distributions for varying alpha values in Fig. 29.3).

Hübner et al. [HBK+11] state that the majority of columns analyzed from financial, sales and distribution modules of an enterprise resource planning (ERP) system were following a power-law distribution—a small set of values that occur frequently, while the majority of values is rare. Furthermore, they identified an average alpha value of 1.581 in enterprise systems.

Figure 29.4 shows the results of an experiment that measures the cumulated log size per query for a varying value distribution. In this experiment, one million INSERT queries of 1,000 distinct zipf distributed values have been simulated. With an alpha value of 1.581, the dictionary is already saturated after $\approx$ 30,000 inserts. Afterwards, queries rarely add entries to the dictionary log. As shown for an alpha value of 4.884 (see Fig. 29.4), the more heavily the distribution is drawn to one value, the smaller the accumulated log size will be. Please note that logging is only needed for queries that modify the data set to save the changes.

A comparison of the log sizes for logical and dictionary-encoded logging is shown in Fig. 29.5. These values have been measured on a productive enterprise system with seven million write operations on the sales item table. Due to the high compression of recurring values, the dictionary-encoded logging reduces the log size by 29 %. As a consequence, dictionary-encoded logging is in favor of logical logging, since it exploits typical data distributions in enterprise systems. For more details about enterprise data characteristics see Chap. 3.

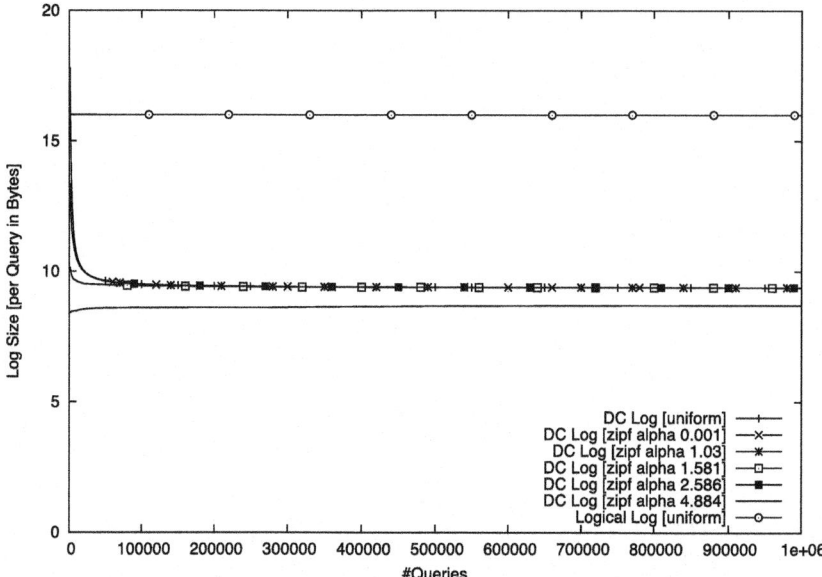

Fig. 29.4 Cumulated average log size per query for varying value distributions (DC = dictionary-compressed)

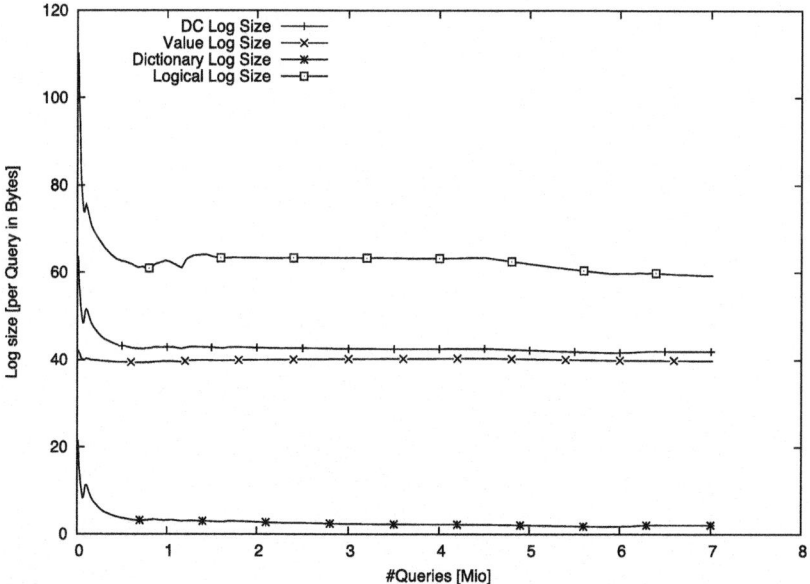

Fig. 29.5 Log size comparison of logical logging and dictionary-encoded logging

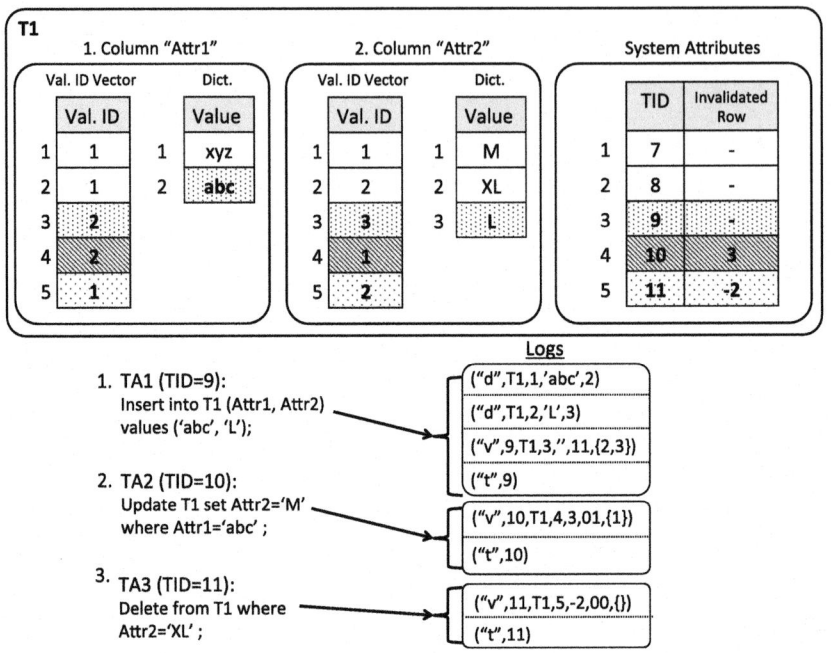

Fig. 29.6 Example: logging for dictionary-encoded columns

29.3 Example

Figure 29.6 shows an example of dictionary-encoded logging. Here, three SQL queries (`insert`, `update`, and `delete`) of three different transactions are logged.

The first statement (`INSERT INTO T1 (Attr1, Attr2) VALUES ('abc', 'L')`) inserts a new row into table T1. This statement has the transaction ID 9 (TID=9), which is stored in the following format:

$$L_t = \{\text{``}t\text{''}, TID\}$$

Since both values ('abc' and 'L') are not yet stored in the corresponding dictionaries, new entries will be added. Dictionary logs L_d are created each time a transaction adds new values into a dictionary. Therefore, the table identifier t, the column index c_i, the added value v, and the corresponding valueID VID are logged. The letter "d" at the first position of the dictionary log entry marks that this is a dictionary log entry, similar to the letter "t" marking the transaction log entry above.

$$L_d = \{\text{``}d\text{''}, t, c_i, v, VID\}$$

Consequently, for transaction 9, two dictionary logs are stored holding the IDs of the modified dictionaries for table T1 (i.e. column ID), the corresponding positions in the dictionaries, and the newly inserted values:

$$(\text{``}d\text{''}, T1, 1, `abc', 2) \; and \; (\text{``}d\text{''}, T1, 2, `L', 3)$$

Value logs L_v store the actual values, which are appended to the attribute vectors. Value log entries store more than just plain data structure changes as done in the dictionary logs, since they have to be linked to the corresponding transactions.

$$L_v = \{\text{``}v\text{''}, TID, t, RID, IRID, bm_n, (VID_1, \ldots, VID_n)\}$$

Each value log L_v hereby stores the transaction identifier TID, the table identifier t, and the rowID RID in the attribute vector. The letter "v" at the first position marks this log entry as a value log entry. The values are stored in a vector of VIDs, whereby the bit mask bm_n marks the corresponding columns (n is the number of attributes in table t). If a row is invalidated by the new row (e.g. due to an update or a delete), the ID of the invalidated row is stored in IRID.

The second statement (UPDATE T1 SET Attr2 = 'M' WHERE Attr1 = 'abc') in Fig. 29.6 alters a row, without introducing new values. Thus only one transactional log entry and one value log entry, storing the new dictionary position for attribute "Attr1", are written to disk.

The delete statement (DELETE FROM T1 WHERE Attr2 = 'XL') invalidates a row. But in this case, the result might not be obvious. When a row is deleted, a new line is added to the table. This is necessary to reflect the changes made by the transaction in the system attributes of the table. In this example, transaction 11 marks row 2 as invalid. This is achieved via hidden system attributes (i.e. columns storing the TID and the corresponding IDs of invalidated rows). While the TID field of a certain row always persists the transaction ID that inserted this row, the invalidated row field is only written for updates and deletes. To mark that a line has not just been updated but entirely deleted, the invalidated row field is prefixed (see entry "−2" in Fig. 29.6). Both, for updates and deletes, the unchanged fields of the inserted tuple are copied from the invalidated row instead of being left empty. The reason is twofold: first, with fixed length attribute vectors, empty fields provide no advantages in terms of performance or memory consumption. Second, copying the row avoids additional lookups to get the values of the invalidated row. This is especially advantageous for long-running transactions and queries, which potentially need to include outdated rows.

It is furthermore important to understand when and in which order the stored logs are written to disk. Once a transaction shall be committed, first the dictionary buffers have to be written to disk. This has to be ensured to avoid value logs referencing valueIDs that cannot be recovered. Afterwards, the value logs are written to disk. Finally, if both logs were written to disk successfully, the committed transaction log is written to disk. Both, the value log entries and the transaction log entries are collected in the same log buffer. This minimizes the amount of flushes to be done upon transaction completion.

To replay the log, the logs are parsed in reverse order (i.e. from the newest to the oldest entry). This is done to ensure that only transactions are written into the differential buffer that have been committed successfully as their commit logs are read before the dictionary and value logs when reading in reverse order.

29.4   Self Test Questions

1. Snapshot Statements
Which statement about snapshots is wrong?

(a) A snapshot is ideally taken after each insert statement
(b) A snapshot is an exact image of a consistent state of the database to a given time
(c) The recovery process is faster when using a snapshot because only log files after the snapshot need to be replayed
(d) The snapshot contains the current read-optimized store

2. Recovery Characteristics
Which of the following choices is a desirable characteristic of any recovery mechanism?

(a) Returning the results in the right sorting order
(b) Recovery of only the latest data
(c) Maximal utilization of system resources
(d) Fast recovery without any data loss

3. Situations for Dictionary-Encoded Logging
When is dictionary-encoded logging superior?

(a) If the number of distinct values is high
(b) If large values are inserted only one time
(c) If all values are different
(d) If large values are inserted multiple times

4. Small Log Size
Which logging method results in the smallest log size?

(a) Log sizes never differ
(b) Logical logging
(c) Common logging
(d) Dictionary-encoded logging

5. Dictionary-Encoded Log Size
Why has dictionary-encoded logging the smaller log size in comparison to logical logging?

(a) Actual log sizes are equal, the smaller size is only a conversion error when calculating the log sizes
(b) Because it stores only the differences of predicted values and real values
(c) Because of the reduction of recurring values
(d) Because of interpolation

References

[Bor84]   A.J. Borr, Robustness to crash in a distributed database: A non shared-memory multi-processor approach, in *VLDB* ed. by U. Dayal, G. Schlageter, L.H. Seng (Morgan Kaufmann, San Francisco, 1984), pp. 445–453

[HBK+11]   F. Hübner, J.-H. Böse, J. Krüger, C. Tosun, A. Zeier, H. Plattner, A cost-aware strategy for merging differential stores in column-oriented in-memory DBMS, in *BIRTE*, pp. 38–52 (2011)

[HR83]   T. Härder, A. Reuter, Principles of transaction-oriented database recovery. ACM Comput. Surv. **15**(4), 287–317 (1983)

[WBR+12] J. Wust, J.-H. Boese, F. Renkes, S. Blessing, J. Krueger, H. Plattner, Efficient logging for enterprise workloads on column-oriented in-memory databases, in *CIKM 2012* (ACM, 2012)

Chapter 30
Recovery

To handle continuously growing volumes of data and intensifying workloads, modern enterprise systems have to scale out, using multiple servers within the enterprise system landscape. With the growing number of servers—and consequently growing number of hard disks and CPUs—the probability of hardware-induced failures is rising.

Productive enterprise systems are expected to never fail or to securely fail-over once a defect is detected. When a server fails, it has to be rebooted and restored, or another server has to take over the workload of the failed server. In either way, to restore the previous state of the server before its failure, data stored on persistent memory has to be loaded back into the in-memory database. This process is called "recovery". Using snapshots and log data—as presented in the previous Chap. 29—a database can be rebuild to the latest consistent state.

The recovery process, which is presented in this section, relies on dictionary-encoded logging [WBR+12]. It is executed in two subsequent tasks: (I) read meta data and prepare data structures, (II) read logging data and recover the database.

30.1 Reading Meta Data

In addition to logging all committed transactions, SanssouciDB logs meta data to speed up the recovery process. With additional knowledge about the data structures, which have to be recovered, expensive data movements and re-allocations can be avoided. Examples for stored meta data are, e.g., the location of the latest snapshot, the number of rows in the main store, or the bits required for the dictionary encoding of each column.

As an example, take the replaying of the dictionary logs. Without knowing in advance how many elements have been persisted before the system failure, the allocated space for the dictionary would probably have to be resized several times. A resize usually requires moving the whole data set to a new allocation. If the number of elements and the number of required bits is known in advance, the allocated space can be sized accordingly without the need of re-allocations or data movements. The number of dictionary elements is stored, since scanning the dictionary log for the latest log to receive the dictionary size is not efficient even when scanning in reverse order. The reason is that transactions do not have to be logged in their incoming order. Thus, finding the maximum dictionary position in the dictionary log might cause reading large parts of the dictionary log file.

30.2 Recovering the Database

After allocating main memory for the data structures, the recovery process continues to replay the database logs. As part of this process, the snapshot of a table's main store is reloaded into memory. At the same time the dictionary log files (containing the dictionary log entries) and the value log files (containing the value and transaction log entries) are replayed.

Due to the dictionary-encoded logging described in Sect. 29.2, the files can be processed in parallel. The import of the dictionary logs and the main store is rather straightforward, while reading the value and transaction log entries from the value log file is a bit more complex. To avoid replaying not committed transactions, the value log file is read in reverse order. This way it is ensured that only value log entries are replayed, whose transactions have been successfully committed. Remember, value and transaction log entries are written to the same file with the strict order of writing the transaction log entry after all value and dictionary log entries have been successfully written.

After the import of the value log file, a second run over the imported tuples is performed. This is caused by the dictionary-encoded logging, which only logs changed attributes of tuples, thus reducing I/O operations. Consequently, the imported tuples have to be checked for empty attributes and they have to be completed if necessary. This is done by iterating over previous versions of the tuple, using the validation flag.

30.3 Self Test Questions

1. Recovery
What is recovery?

   (a) It is the process of recording all data during the run time of a system
   (b) It is the process of cleaning up main memory, to "recover" space
   (c) It is the process of restoring a server to the last consistent state before a crash
   (d) It is the process of improving the physical layout of database tables to speed up queries

2. Server Failure

What happens in the situation of a server failure?

(a) The failure of a server has no impact whatsoever on the workload
(b) The system has to be rebooted and restored if possible, while another server takes over the workload
(c) The power supply is switched to backup power supply so the data within the main memory of the server is not lost
(d) All data is saved to persistent storage in the last moment before the server shuts down

Reference

[WBR⁺12] J. Wust, J.-H. Boese, F. Renkes, S. Blessing, J. Krueger, H. Plattner, Efficient logging for enterprise workloads on column-oriented in-memory databases, in *CIKM 2012* (ACM, 2012)

Chapter 31
Database Views

Views provide means to store queries as virtual tables (also logical tables) [ABC⁺76]. They define a structured subset of the data available in the database. Views do not store data, but describe a transformation rule on the underlying data that is processed when the view is accessed [PZ11].

31.1 Advantages of Views

The concept of database views comes with two major advantages.

First, they can be used to reduce the complexity of queries, e.g., data type conversions, special transformations and joins are hidden by the view while the transformed data is returned as result. Multiple views can be cascaded. Thus, complex queries can be orchestrated and better maintained.

Second, views replace long-running transformations to integrate other data sources currently performed via ETL processes by instant transformations. For example, if a data transformation is required, an ETL process requires to transform all data before importing them. With the help of views, the transformation is performed just when a certain data item is accessed. This is advantageous, if only a small subset of all imported data is accessed.

Figure 31.1 shows the view `metropolises_population`, which returns only citizens from cities with more than one million inhabitants. With the help of this view, writing queries that select citizens of big cities is straightforward while the readability and understandability of the query is improved.

Views can also be used to implement parts of the security requirements for applications. User and access management for example are aspects than can be handled via views [BH88, Wil88].

Another advantage of database views is that they can be used to create virtual data schemas that build stable interfaces for application development. Two important aspects concerning software quality, software maintenance and reusability, are enhanced by the decoupling of the application code from the actual data

H. Plattner, *A Course in In-Memory Data Management*,
DOI 10.1007/978-3-642-55270-0_31, © Springer-Verlag Berlin Heidelberg 2014

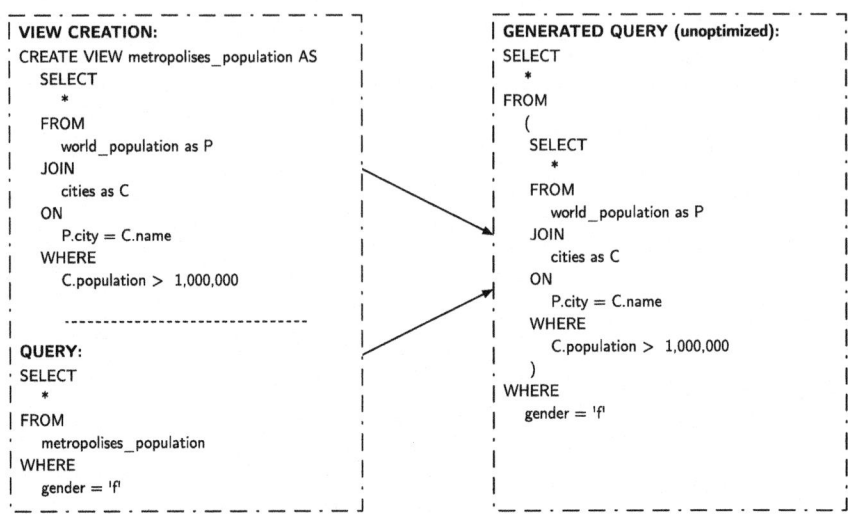

Fig. 31.1 Using views to simplify join-queries

schema. A prominent example are data cubes, such as those used in many data warehouses [GBLP96].

Instead of materializing data redundantly as a cube schema, virtual cubes can be created on the fly with the help of database views. In contrast to traditional cubes, virtual cubes work directly on the raw data, i.e. they access always the latest data without any latency. As a result, the integration of third party software applications, such as Business Intelligence dashboards, Microsoft Excel, or web applications, is simplified, storage demands are reduced due to the elimination of redundant data, and forecasts become more accurate since data used by virtual cubes is always the latest available data within the database.

Some of the aforementioned advantages, such as the possibility to transform and filter data or hide joins cause limitations for modifying operations. Therefore, inserts and updates into views are usually not possible, since there is no bijective mapping between the presented entries in the view and the originating entries in the underlying tables.

31.2 Layered Views Concept

Figure 31.2 depicts the *layered view concept*. This concept describes the assembly of views in layers. Hereby, the data sources for a view can be either tables or other views. Views can be built on column-oriented and row-oriented database tables equally. The layered view concept allows the integration of external data sources, such as other databases, to join them into one virtual table. Thus, the layered view concept can simplify the development of queries and the combination of data.

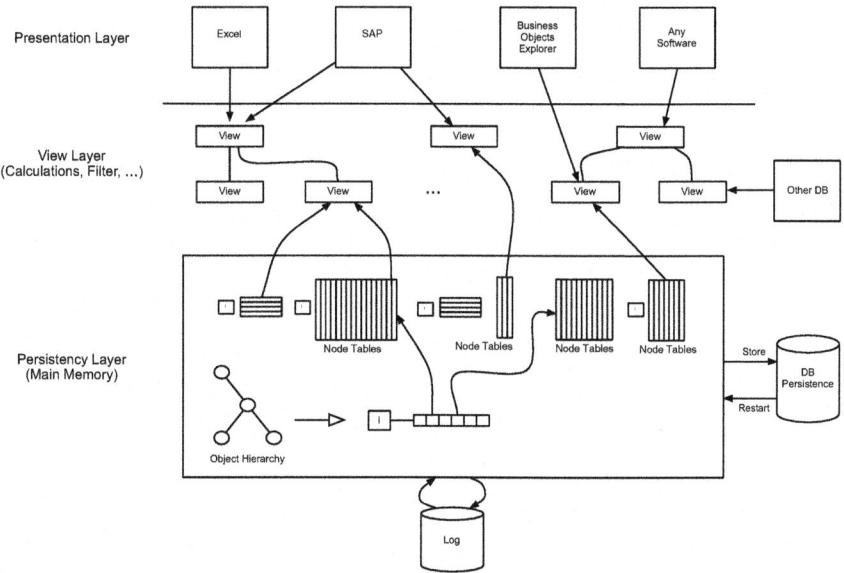

Fig. 31.2 The view layer concept

Graphical tools can be used for view creation. These tools are able to create complex join-views by interactively dragging one database table onto another, whereby join attribute(s) are automatically determined. Furthermore, view development tools provide performance analysis, e.g., of a joined table, and point out possible processing improvements by rearranging data or optimized query execution plans.

31.3 Self Test Questions

1. View Locations
Where should a logical view be built to get the best performance?

(a) close to the user in the analytical application
(b) in a third system
(c) in the GPU
(d) close to the data in the database

2. Views and Software Quality
Which aspects concerning software quality are improved by the introduction of database views?

(a) Accessibility and availability
(b) Testability and security
(c) Reusability and maintainability
(d) Fault tolerance and usability

References

[ABC⁺76]   M.M. Astrahan, M.W. Blasgen, D.D. Chamberlin, K.P. Eswaran, J.N. Gray,
           P.P. Griffiths, W.F. King, R.A. Lorie, P.R. McJones, J.W. Mehl, G.R. Putzolu,
           I.L. Traiger, B.W. Wade, V. Watson, System r: relational approach to database
           management. ACM Trans. Database Syst. **1**(2), 97–137 (1976)
[BH88]   E. Bertino, L.M. Haas, Views and security in distributed database management
           systems, in ed. by J.W. Schmidt, S. Ceri, M. Missikoff, *Advances in Database
           Technology'EDBT '88*, number 303 in Lecture Notes in Computer Science, pp. 155–
           169 (Springer, Berlin/Heidelberg, 1988)
[GBLP96]   J. Gray, A. Bosworth, A. Layman, H. Pirahesh, Data cube: A relational aggregation
           operator generalizing group-by, cross-tab, and sub-total, in *ICDE* ed. by S.Y.W. Su
           (IEEE Computer Society, Washington, DC, 1996), pp. 152–159
[PZ11]   H. Plattner, A. Zeier, *In-Memory Data Management* (Springer, Heidelberg, 2011)
[Wil88]   J. Wilson, Views as the security objects in a multilevel secure relational database
           management system, in *Proceedings of the 1988 IEEE Conference on Security and
           Privacy*, SP'88 (IEEE Computer Society, Washington, DC, 1988), pp. 70–84

Chapter 32
On-the-Fly Database Reorganization

In typical enterprise applications, schema and data layout have to be changed from time to time. The main cases for such changes are software upgrades, software customization, or workload changes. Therefore, the option of database reorganization, such as adding an attribute to a table or changing attribute properties, is required.

In row-oriented databases, database reorganization is typically time-consuming as well as cost-intensive. That is why most row-oriented database management systems usually do not allow data definition operations while the database is online [AGJ⁺08]. Consequently, downtime of the database server has to be taken into account. In contrast, modifications within a column store database, such as SanssouciDB, can be done dynamically without any downtime. The following sections explain what database reorganization looks like in row stores and column stores.

32.1 Reorganization in a Row Store

In row stores, database reorganization is expensive. As mentioned in Sect. 8.2 all attributes of a tuple are stored sequentially in the same memory block, where each block contains multiple rows. The left side of Fig. 32.1 shows a table that includes a unique identifier, the first name, and last name of citizens.

If an additional attribute, for example, state is added and no space is available in the block, adding a new attribute requires a reorganization of the storage for the entire table. The same issue occurs when the size of an attribute is increased. The right side of Fig. 32.1 shows the table's storage after the attribute *state* is added. Each row is extended by that attribute and all following rows are moved within the block (and the following blocks if necessary).

To be able to dynamically change the data layout, a common approach for row stores is to create a logical schema on top of the physical data layout [AGJ⁺08]. This method allows changing the logical schema without modifying the physical

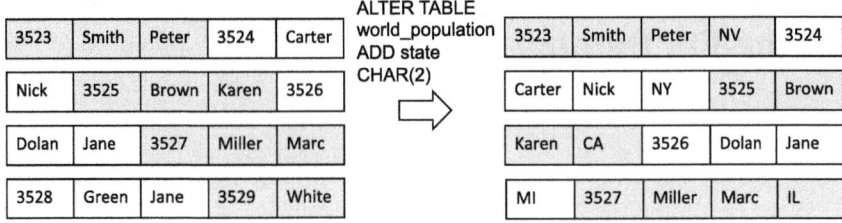

Fig. 32.1 Example memory layout for a row store

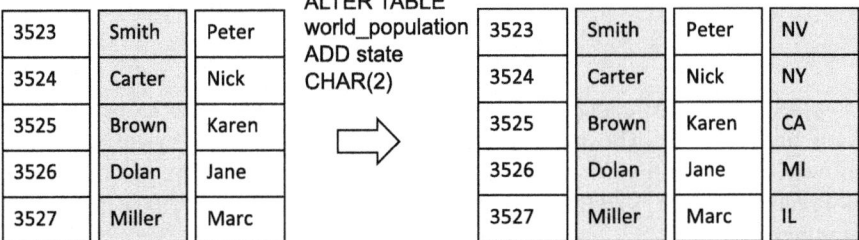

Fig. 32.2 Example memory layout for a column store

representation of the database but also decreases the performance because of the overhead of accessing the meta data and data of the logical tables. Another approach is using schema versioning for database systems [Rod95]. These advanced approaches will not be discussed any further as part of this learning material.

32.2   On-the-Fly Reorganization in a Column Store

In column-oriented databases, each column is stored independently from the other columns in a separate block, see the example in Fig. 32.2. New attributes can be added very easily, because they will be created in a new memory area. Locking for changing the data layout is only required for a very short period, during which solely the meta data of the table is adapted.

In SanssouciDB data structures for new columns will not be allocated before the first value is added. The dictionary and the attribute vector of new columns remain non-existent as long as the column does not contain any values. The addition of a column has no impact whatsoever on existing applications if they solely request their required attributes from the database (meaning they do not use *SELECT* * statements).

32.3 Excursion: Multi-Tenancy Requires Online Reorganization

This sections gives a typical use case where online reorganization is required in a database system.

In a *single-tenant* system, each customer (tenant) has its own database instance on a physically separated server machine. In this case, maintenance costs for the service provider will be very high and in addition, tenants do not even use their system permanently with the complete utilization level. In contrast, on *multi-tenant* systems different customers share the same resources on the same machine. By providing a single administration framework for the whole system, multi-tenancy can improve the efficiency of system management [JA07] and increases the utilization of systems. The Software-as-a-Service provider Salesforce.com[1] first employed this technique on a large scale.

Multi-tenancy can be implemented in three different ways with different levels of granularity: shared machine, shared database instance, and shared table.

In the *shared machine* implementation (see Fig. 32.3a) each customer has its own database process and these processes are executed on the same server. The advantages of this approach are a good isolation among other tenants and easy customer migrations from one machine to another. Major limitations are that this approach does not support memory pooling and each database needs its own connection pool. Moreover, administrative operations cannot be applied on all database instances simultaneously (in bulk).

In the *shared database instance* implementation (see Fig. 32.3b) each customer has its own tables, but shares the database instance with other customers. In this case, connection pools can be shared between customers and pooling of memory is better compared with the previous approach. On the other hand, isolation between customers is reduced. This approach allows simultaneous execution of many administrative operations on all database instances.

In the *shared table* approach (see Fig. 32.3c) many tenants share the common database and each customer has its own rows, which are marked by an additional attribute, e.g., *tenantID*. With this approach, resource pooling performs best and sharing of connection pools between customers is possible. Administrative operations can be carried out in bulk by running queries over the column containing the *tenantID*.

Multi-tenant systems using the shared table approach are a typical environment where on-the-fly database reorganization is necessary. These systems aim at maintaining the ability of individual tenants to make custom changes to their database tables, while not affecting other tenants using the same resources. In row stores, the entire database or table would be completely locked to process data definition operations. In the column store, the table is locked only for the amount of time that

---

[1]http://www.salesforce.com.

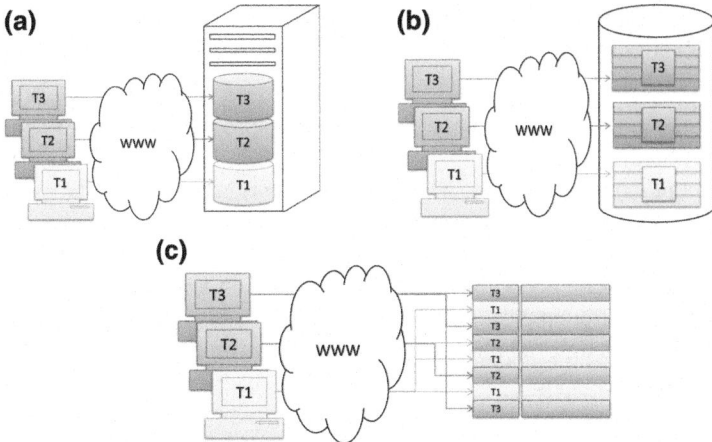

Fig. 32.3 Multi-tenancy granularity levels. (**a**) Shared machine. (**b**) Shared database instance. (**c**) Shared table

is needed to complete the data definition operation and the lock is only restricted to the meta data of the table.

32.4   Hot and Cold Data

In addition to an ever-increasing growth of data, internal (e.g. controlling) and external (e.g. tax law) requirements demand that data is kept at hand for many years. A separation of data, based on its application relevance into *cold data* and *hot data* is an approach to improve memory utilization and handle the increasing, but still finite capacity of main memory as efficiently as possible. The ability to exclude cold data for a majority of query executions, while still returning correct results in cases cold data is needed, also promises considerable performance improvements.

The idea of separating hot and cold data is based on the observation of different data access frequencies and skewness, especially for transactional tables that usually account for a good portion of the overall data volume. It is typical for enterprise applications that major parts of the workload request only a small, recurring fraction of the overall data. While certain tuples are often selected, others are seldom or even never part of a query result set. Such less relevant data should be handled with lower priority than highly relevant data. For example in case of DRAM shortage, a relevance-based classification contributes to better decisions on what data can be unloaded. This improves system behavior when the data volume grows beyond the available main memory, concurrent transactions use up all memory , or when a particularly complex query performs a cross-join on very large tables. In the end we have a more efficient database system that can handle growing data volumes or a higher workload using the same resources.

32.4.1   Data Classification and Access

There are two main aspects to this challenge: *Data Classification* and *Data Access*. Data classification concepts can be separated based on their data granularity (table-, column-, block-, tuple-, or range-wise), frequency (online or offline), and level (database or application) of classification.

Data access handling can be done on application as well as on database level. There are explicit and implicit access concepts. Explicit means that the type of access (hot or cold) is known before a query is executed, while implicit ones rely on separate layers for handling access (e.g. buffer manager). Every concept that classifies data also has to consider how and when which data needs to be accessed, because only if access to less optimized data can be avoided or reduced during query execution, data differentiation makes sense.

There are two existing, fundamentally different approaches that are presented in the next two sections. Either, the application layer controls it (Sect. 32.4.2), harnessing data and process-specific characteristics, or the database does it transparently to the application (Sect. 32.4.3). Finally in Sect. 32.4.4, the new approach in SanssouciDB is described.

32.4.2   Application-Driven Classification

Existing approaches on application level use complex rule sets that scan and classify data tuple-wise, using application, customer, and data knowledge. As described by Mensching et al. in [MC04], such approaches are used to classify archivable table data in enterprise systems. For performance reasons, but also from a legal point of view, the archive is by default excluded from query execution. The application must have knowledge of the archiving rules, so it can explicitly tell the database when a query (data selection) has to access archived data as well in order to return correct query results. This is manageable as long as there are only few queries and well-defined application access points for archived data. However, if we want to classify more data as not relevant the application complexity of handling hot and cold data access rises.

Another approach is built on consistent object models that map business application objects to database tables. As described in [GKM+11], most business objects in enterprise applications have a life cycle, similar to the sales opportunity shown in Fig. 32.4. The life cycle of a business object can be separated into active and passive states. While passive data is typically read-only, active data is being frequently processed and can consequently still be modified. Classifying database tuples based on the state of their related application object implicitly harnesses process information while establishing a contract between application and database. This results in a strong dependency between application logic and elementary database specific storage mechanisms. Such a behavior is not always desirable. Especially for

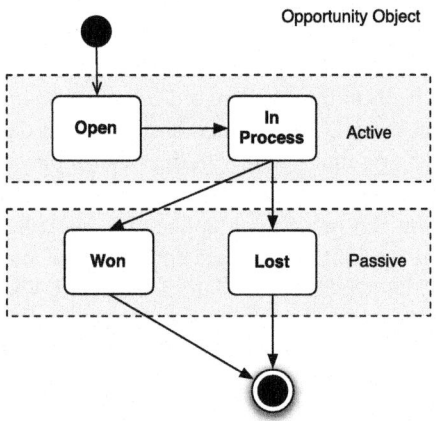

Fig. 32.4 The life cycle of a sales order

enterprise applications with various data access points and many customer specific
extensions, it seems questionable whether a classification-based approach using
such a contract provides the necessary flexibility. Analytical queries for example
usually do not adhere to defined object states. If data is relevant for an application
workload, the data's relevance does not depend on its object state alone. For example
lost or won business opportunities can still be relevant for certain business scenarios
and user groups.

32.4.3   Database-Driven Classification

The most prominent concepts of handling data relevance on the database layer
are caching algorithms such as Least Recently Used (LRU) or Least Frequently
Used (LFU). Optimized for disk access as their primary storage layer, disk-based
databases rely on caching of pages that typically contain a block of 4 kilobyte
data. A buffer manager then uses algorithms such as LRU and LFU to keep hot
pages cached. It tracks the access to frequently and recently used pages so that
most requests can be answered without disk IO. However, for performance reasons,
this layer of indirection is dropped from in-memory databases so that data is byte-
addressable (see Sect. 5.1). Data structures, such as main or delta store can be
accessed directly without checking if the requested page is cached or not.

32.4.4   Application Transparent Classification

The concept of database-driven classification is not applicable for in-memory
databases for two reasons. First, there is no need of splitting data into disk pages

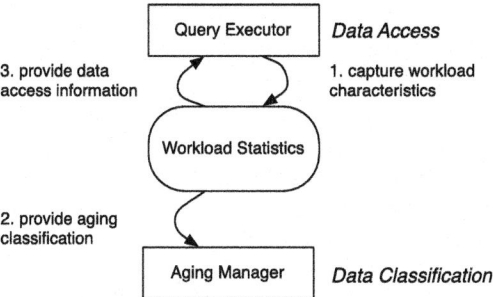

Fig. 32.5 Data classification and access

anymore. Data is considered to be byte-addressable and stored in main memory so no cache checking is required. Second, as there are mixed workloads in modern enterprise applications, neither a sole horizontal nor a sole vertical packaging of data into blocks is an adequate solution.

Still, as illustrated in Fig. 32.5, data classification as well as data access in SanssouciDB are application transparent. SQL statements of the application workload and their data selection characteristics are captured asynchronously during query execution. These *Workload Statistics* are used by the *Tiering Manager* in order to classify entire tuples or columns during an tiering run into hot and cold partitions. After classifying data the *Workload Statistics* provide data access information to the *Query Executor*, used to determine queries that can be executed on hot data only. Others, that do not match the *Workload Statistics* have to consider cold data as well during execution. The tiering run is started automatically as required, running asynchronously to the query processing. Having a differential buffer architecture with delta and main store in SanssouciDB, the merge process is an adequate point of time to evaluate data relevance. However, a tiering run can also be triggered when the rate of hot-only queries goes below a defined threshold or the size of the hot data partition exceeds a maximum.

Query Executor (data access) and *Tiering Manager* (data classification) operate on the same statistics, extracted from the SQL workload. This way, the relevance of data is determined based on the recent application workload, instead of fixed data rules. The number of queries that run on the hot data partition is optimized, while at the same time the memory size of hot data is reduced. There are no dependencies between application and database necessary to provide transparent tiering between different logical storage classes of an in-memory database. Application transparent classification simplifies the design of the overall application, while transparent data access also disposes the database administrator and users from the difficult decision which data to tier. Data moved to less performant storage tiers (*cold data*) remains accessible the same way as *hot data*.

Workload Statistics consist of selection clauses and selection values. *Selection clauses* are all distinct WHERE-clauses used to select data from a table (e.g. ColA = ? AND ColB > ?). The number of distinct selection clauses is condensed by eliminating attributes that do not indicate data relevance (e.g. gender). For each selection clause, the bind attributes (called *selection values*) are captured during query execution (e.g. ColA = 1 AND ColB > 14-01-03). Data is classified during a tiering run, using selection clauses as well as the usage frequency and patterns of the selection values .

Once the data is classified, it can be treated with different priorities. Different data structures, compression techniques, storage media (DRAM for hot data, SSD for cold data), and materialization strategies can then further be used.

32.5   Self Test Questions

1. **Separation of Hot and Cold Data**
   How should the data separation into hot and cold take place?

   (a) Application transparently, depending on workload relevance
   (b) Automatically, depending on the state of the application object in its life cycle
   (c) Block-wise, because data is already structured into horizontal blocks
   (d) Manually, upon the end of the life cycle of an object

2. **Data Reorganization in Row Stores**
   The addition of a new attribute within a table that is stored in row-oriented format . . .

   (a) is very cheap, as only meta data has to be adapted
   (b) is possible on-the-fly, without any restrictions of queries running concurrently that use the table
   (c) is an expensive operation as the complete table has to be reconstructed to make place for the additional attribute in each row
   (d) is not possible

3. **Cold Data**
   What is cold data?

   (a) Data, which is still accessed frequently but is not updatable any longer
   (b) Data that is used in a majority of queries
   (c) The rest of the data within the database, which does not belong to the result of the current query
   (d) Data that is seldom or even never returned in query results

4. **Data Reorganization**
   The addition of an attribute in the column store . . .

   (a) slows down the response time of applications that only request the attributes they need from the database
   (b) has no impact on existing applications if they only request the attributes they need from the database

(c) speeds up the response time of applications that always request all possible attributes from the database

(d) has no impact on applications that always request all possible attributes from the table

5. Single-Tenancy

In a single-tenant system ...

(a) power consumption per customer is best and therefore it should be favored

(b) all customers are placed on one single shared server and they also share one single database instance

(c) each tenant has its own database instance on a shared server

(d) each tenant has its own database instance on a physically separated server

6. Shared Machine

In the shared machine implementation of multi-tenancy ...

(a) each tenant has an own exclusive machine, but these share their resources (CPU, RAM) and their data via a network

(b) all tenants share one server machine, but have own database processes

(c) all tenants share the same physical machine, but the CPU cores are exclusively assigned to the tenants

(d) each tenant has an own exclusive machine, but these share their resources (CPU, RAM) but not their data via a network

7. Shared Database Instance

In the shared database instance implementation of multi-tenancy ...

(a) each tenant has its own server, but the database instance is shared between the tenants via an InfiniBand network

(b) all tenants share one server machine and one main database process, tables are also shared

(c) the risk of failures is minimized because more technical staff (from different tenants) will have a look at the shared database

(d) all tenants share one server machine and one main database process, tables are tenant exclusive, access control is managed within the database

References

[AGJ+08]   S. Aulbach, T. Grust, D. Jacobs, A. Kemper, J. Rittinger, Multi-tenant databases for software as a service: schema-mapping techniques, in *Proceedings of the International Conference on Management of Data*, SIGMOD '08 (ACM, New York, 2008), pp. 1195–1206

[GKM+11]   M. Grund, J. Krueger, J. Mueller, A. Zeier, H. Plattner, Dynamic partitioning for enterprise applications, in *MTT-S International Microwave Workshop Series on Innovative Wireless Power Transmission: Technologies, Systems, and Applications* (IEEE, 2011), pp. 1010–1015

[JA07]   D. Jacobs, S. Aulbach, Ruminations on multi-tenant databases, in *BTW* ed. by
           A. Kemper, H. Schöning, T. Rose, M. Jarke, T. Seidl, C. Quix, C. Brochhaus, vol. 103
           of *LNI* (GI, Aachen, 2007), pp. 514–521
[MC04]   J. Mensching, G. Corbitt, ERP data archiving - a critical analysis. J. Enterprise
           Inform. Manag. **17**(2), 131–141 (2004)
[Rod95]   J.F. Roddick, A survey of schema versioning issues for database systems. Inform.
           Software Tech. **37**(7), 383–393 (1995)

Part V
Principles for Enterprise Application Development

Chapter 33
Implications on Application Development

In the previous chapters, we introduced the ideas behind our new database architecture and their technical details. In addition, we showed that the in-memory approach can significantly improve the performance of existing database applications.

In this chapter, we discuss how the existing applications should be redesigned and how new applications should be designed to take full advantage of the new database technology. Our research and the prototypes we built show that in-memory technology greatly influences the design and development of enterprise applications. The main driver for these changes is the drastically reduced response time for database queries. Now, even more complex analytical queries can be executed directly on the transactional data in less than one second. With this performance, we are able to develop new applications and enhance currently existing applications in a way that was not possible before. Modern applications can especially benefit from the database performance when it comes to better granularity and actuality of the processed data.

The most important approach to achieve this performance is to move application logic closer to the database. While traditional approaches try to encapsulate complex logic in the application server, with the advent of in-memory computing it becomes crucial to move data intensive logic as close as possible to the database. An additional advantage of moving data-intensive application logic closer to the database is that the amount of data that has to be transferred between the application server and the database system is significantly reduced when most of the data intensive operations are executed directly in the database system.

33.1 Optimizing Application Development for In-Memory Databases

A typical enterprise application contains three main architectural layers as shown in Fig. 33.1.

H. Plattner, *A Course in In-Memory Data Management*,
DOI 10.1007/978-3-642-55270-0_33, © Springer-Verlag Berlin Heidelberg 2014

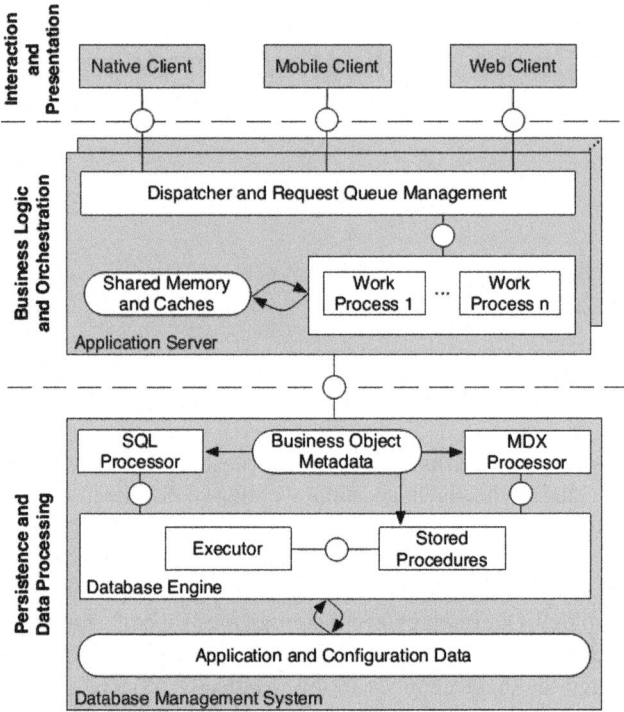

Fig. 33.1 Three layer enterprise application

These three main layers are usually distributed over three independent physical systems, which leads to a three tier setup. To ensure a common understanding of the terms layer and tier, the words are shortly explained: A *layer* separates program code and its responsibility on the logical level, but it does not state how the deployment of the code looks like. The word *tier* describes the physical architecture of a system, so it gives details about the hardware setup used to run the program code.

The interaction and presentation layer is responsible for providing a user interface. Moreover, the presentation layer gets user information requests and forwards them to the underlaying layers. The complete user interface may consist of many different independent parts for different devices or platforms.

The business logic and orchestration layer acts as a mediator between the presentation and the persistence layer. It handles user requests obtained from the presentation layer. This can either be the direct execution of data operations by using the application's cache or delegation of calls to the persistence layer.

Data persistence and processing provides interfaces for requesting data with the help of declarative query languages, such as SQL or Multidimensional Expressions (MDX), and prepares data for further processing in the upper layers.

```
for customer in allCustomers () do
  for invoice in customer.unpaidInvoices () do
    if invoice.dueDate < Date.today()
      dueInvoiceVolume [customer.id] +=
        invoice.totalAmount
    end
  end
end
```

Listing 33.1 Imperative implementation (pseudo code) of a dunning run

33.1.1 Moving Business Logic into the Database

As mentioned before, in traditional applications the application logic is mainly stored in the orchestration layer to allow easier scaling of the complete application and to reduce the load of the database. To leverage the full performance, we have to identify which application logic should be moved closer to the persistence layer. The ultimate goal is to leave only such logic in the orchestration layer that provides functionality that is orthogonal to what can be handled within the user interaction request. This reduced layer would then mostly translate user requests into SQL and MDX queries, or calls to stored procedures on the database system.

To illustrate the impact, we will explain the changes and the effects using an example that performs an analytical operation directly on the transactional data. In the following, two different implementations of the same user request will be compared. The request identifies all due invoices per customer and aggregates their amount. This is usually referred to as dunning and is one of the most important applications for consumer companies. It is typically a very time-consuming task, because it involves read operations on large amounts of transactional data.

Listing 33.1 implements business logic directly in the application layer. It depends on given object structures and encodes the algorithms in terms of the used programming language.

Using this approach, all customer data is required to be loaded from the database and an object instance for each customer will be created. To create the object, all attributes will be loaded although only one attribute is needed. After that, for all invoices of each customer it will be determined whether it is considered paid or not. For that, it is checked whether the due date at which the invoice should be paid has already passed. Finally, the total unpaid amount for each customer is aggregated.

For each iteration of the inner loop, a query is executed in the database to retrieve all attributes of the customer invoice. This causes bad runtime performance.

The second approach, presented in Listing 33.2, uses a single SQL query to retrieve the same result set. All calculations, filtering and aggregations are handled close to the data. Therefore, the efficient and parallelized operator implementation introduced in previous chapters can be used. The other advantage is that only the required result set is returned to the application layer. Consequently, network traffic is reduced.

```
SELECT invoices.customerId,
  SUM(invoices.totalAmount) AS dueInvoiceVolume
FROM invoices
WHERE invoices.isPaid IS FALSE
  AND invoices.dueDate < CURDATE()
GROUP BY invoices.customerId
```

Listing 33.2 Declarative implementation of a dunning run in SQL

When using small amounts of data, the performance differences are barely noticeable. However, once the system is in production mode and filled with realistic amounts of data, using the imperative approach results in much slower response times. Accordingly, it is very important to test performance of different algorithms with realistic customer data sets that represent realistic sizing settings and value distributions.

The ability to express application logic using SQL can be a huge advantage because expensive calculations are done inside the database. That way, calculations as well as comparisons can work directly on the compressed data. Only as the last step, when returning the results, the compressed values are converted to the original values to present them in human readable format.

33.1.2 Stored Procedures

An additional possibility to move application logic into the database are *stored procedures* which allow to reuse data-intensive application logic. The main benefits of using stored procedures are:

- Business logic centralization and reuse,
- Reduction of application code and simplification of change management,
- Reduction of network traffic, and
- Pre-compilation of queries increasing the performance for repeated execution.

Stored procedures are typically written in a special mixed imperative-declarative programming language as shown in Listing 33.3. Such programming languages support both declarative database queries, e.g. SQL, and imperative controlling sequences, such as loops or conditions, and concepts, e.g. variables or parameters. Once a stored procedure is defined, it can be used and reused by several applications. Applicability across different applications is usually established via individual invocation parameters. Our tiny example does not contain such parameters, but we could alter it so that we pass a country to the procedure which is used as a selection criterion and only customers of this country would be part of the dunning run.

```
// Definition
CREATE PROCEDURE dueInvoiceVolumePerCustomer ()
BEGIN
  SELECT invoices.customerId ,
    SUM( invoices.totalAmount ) AS dueInvoiceVolume
  FROM invoices
  WHERE invoices.isPaid IS FALSE
    AND invoices.dueDate < CURDATE()
  GROUP BY invoices.customerId
END

// Invocation
CALL dueInvoiceVolumePerCustomer ()
```

Listing 33.3   Creation of a stored procedure

33.1.3   Example Application

One prominent example, where we were able to achieve an astonishing performance increase over a traditional implementation was in the area of financial applications. Here, we analyzed the dunning run, meaning extraction of all overdue accounting entries from the accounting tables. The traditional picture of the dunning run showed that the original application was implemented as follows: First, all accounts that have to be dunned are selected and transferred as a list to the application server. Second, for each account all open account items are selected and the due date for each of them calculated. Third, for all items to be dunned, additional configuration logic is loaded and the materialized result set written to a dedicated dunning table. From the discussion in the previous sections we see that this implementation is clearly disadvantageous since it executes a lot of individual SQL statements and transfers intermediate results from the database system to the application server and back. In addition, the implementation looks like a manual join implementation connecting accounts with account items.

In several iterations on the dunning implementation, we were able to reduce the overall runtime of the dunning implementation from initially $1,200s$ to $1.5s$. Figure 33.2 shows the summary comparison of these implementations. The main difference between the versions is that the fastest implementation tries to push as much selection already down to the first filter predicates and executes as much as possible in parallel. Thus, we were able to achieve a speedup of factor 800.

To summarize, in our new implementation of the dunning run we followed the principles that have been presented earlier in this section. The most important of these principles is to move data intensive application logic as close as possible to the database.

Original Version needed about 20 minutes
→ Factor 800x acceleration achieved

#	Operation	HANA2 Version	Variant 2	Variant 3
1	Select Open Items	0.63s	1.01s (incl. T047 & KNB5 Join)	0.6s (incl. T047 & KNB5 Join)
2	Due date, dunning level	27s	deferred to aggregation	0.5s
3	Filter 1 (Verify Dunning levels)	≈ 19s	1.1s	0.5s
4	Filter 2 (Check Last Dunning)	≈ 15s	0.8s	0.4s
5	Generate MHNK (Aggregate)	done in #1	1.2s	done in #1
6	Generate MHND (Execute Filters)	done in #1	140ms	done in #1
	Total	≈ 1 Minute	≈ 3.0s (#3, #4 exec. in parallel)	≈ 1.5s (#3, #4 exec. in parallel)

Fig. 33.2 Comparison of different dunning implementations

33.2   Best Practices

In the following section, the discussion of Chap. 33 will be summarized by outlining the most important rules which should be followed when developing enterprise applications.

- *The right place for data processing:* This is an important decision, which developers have to make during implementation. The more data is processed during a single operation, the closer it should be executed to the database. Aggregations should be executed in the database while single record operations should be part of the application layer.
- *Avoid SELECT *:* Only really required attributes for the application should be loaded. Developers often tend to load more data than is actually needed, because this apparently allows easier adoption to unforeseen use cases. The downside is that this leads to intensive data transfer between the application and database servers which causes significant performance penalties. Furthermore, tuple reconstruction in a column-oriented data format is slightly more complex than in a row-oriented data format as shown in Chap. 13.
- *Use real data for application development:* Only real data can show possible bottlenecks of the application architecture and identify patterns that may have a negative impact on the application performance. Another benefit is that user feedback during development tends to be much more productive and precise if real data is used.
- *Work in inter-disciplinary teams:* We believe that only joint, multidisciplinary efforts of user interface designers, application programmers, database specialists, and domain experts will lead to the creation of new, innovative applications. Each of them has its own point of view and is able to optimize one aspect of a possible

solution, but only if they jointly try to solve problems, the others will benefit from their knowledge.

33.3 Self Test Questions

1. Architecture of a Banking Solution
Current financial solutions contain base tables, change history, materialized aggregates, reporting cubes, indices, and materialized views. The target financial solutions contains ...

(a) only indexes, change history, and materialized views.
(b) only base tables, algorithms, and some indexes.
(c) only base tables, materialized aggregates, and materialized views.
(d) only base tables, reporting cubes, and the change history.

2. Criterion for Dunning
What is the criterion to send out dunning letters?

(a) When the responsible accounting clerk has to achieve his rate of dunning letters
(b) Bad stock-market price of the own company
(c) A customer payment is overdue
(d) Bad information about the customer is received from consumer reporting agencies

3. In-Memory Database for Financials
Why is it beneficial to use in-memory databases for financials systems?

(a) Because of the high reliability of data in main memory, less maintenance work is necessary and labor costs could be reduced.
(b) Easier algorithms are used within the applications, so shorter algorithm run time leads to more work for the end user. Business efficiency is improved.
(c) Operations like dunning can be performed in much shorter time.
(d) Financial systems are usually running on mainframes. No speed up is needed. All long-running operations are conducted as batch jobs.

4. Languages for Stored Procedures
Languages for stored procedures are ...

(a) strongly declarative, they just describe how the result set should look like. All aggregations and join predicates are automatically retrieved from the database, which has the information "stored" for that.
(b) strongly imperative, the database is forced to exactly fulfill the orders expressed via the procedure.
(c) usually a mixture of declarative and imperative concepts.
(d) designed primarily to be human readable. They follow the spoken English grammar as close as possible.

Chapter 34
Handling Business Objects

The notion of objects as a means to structure code reaches back to the 1960s when Dahl and Nygaard invented Simula-67 [DN66]. Object-oriented programming as a new programming paradigm was introduced by Alan Kay with the Smalltalk programming language in the 1970s [Ing78]. Since then, object-orientation has evolved into the dominant programming paradigm for applications in various domains. Especially enterprise applications with their inherent aim to capture properties, behavior, and processes of real world companies benefit from object-oriented programming features. Concepts such as encapsulation, aggregation, and inheritance provide system architects with the means to design domain models, which reflect structures and relations of the real world. Based on such domain models, developers are able to communicate and discuss business logic with domain experts to verify conceptual and logical correctness of the system's functionality.

In the context of object-oriented programming languages, classes or objects are a means to structure code and to enforce the paradigm of the separation of concerns. Based on that definition, an object can represent a connection to a database, a button for the user interface, a file in the file system, or simply a list of other objects. In the context of this book, we narrow the notion of a object in the sense that we consider objects to be a technical or logical entity that has a semantic meaning in the context of a business. We call them business objects. Prominent examples of such objects are a *customer*, an *address*, a *sales order*, a *bank*, or an *invoice*. Typically, business objects have a tree-like structure where the tree's nodes contain data or associations to other business objects. Figure 34.1 depicts the structure of a *sales order*.

It consists of a *Header* with general information like the order number, the order date and the customer (business partner), a collection of *Delivery Terms*, and a collection of *Items*. Every *Item* consists of a collection of *ScheduleLines*. The presence of sub-structures (e.g., *Delivery Terms* in our example) may be optional. The reason for that is twofold. For ones, the relevant information for that sub-structure might simply not have been added to the system, yet (e.g., the customer has already added the items of his order to the shopping cart, but she did not yet decide when and how to ship). The second and more complex reason is that

H. Plattner, *A Course in In-Memory Data Management*,   235
DOI 10.1007/978-3-642-55270-0_34, © Springer-Verlag Berlin Heidelberg 2014

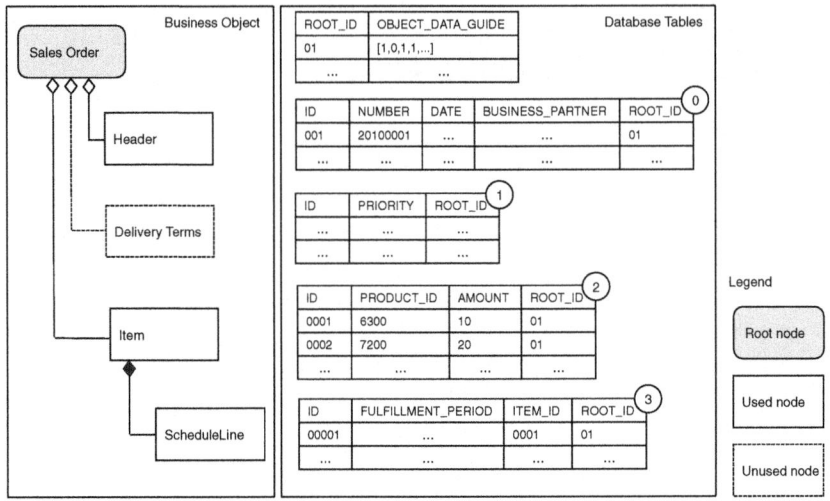

Fig. 34.1 Sales order business object with object data guide representation

not every piece of functionality of an enterprise information system is used by every company. Companies are very specific in terms of shape and extent of their business models and consequently so are the business processes they implement. The solution to that situation is called customization. It is a major feature in every enterprise application. One way to accommodate customization possibilities is to offer comprehensive business objects, which are able to capture every aspect of a business process or business entity. However, most companies will not require every aspect of a business object. Thus, sub-structures of business objects may be optional. Looking at Fig. 34.1, we can see that the complex object *sales order* is stored across several tables. Retrieving such an object form the database requires a number of joins. Depending on whether or not an optional sub-structures is filled determines the number of necessary joins that need to be executed to fetch the business object from the database. To avoid unnecessary joins, we introduce the concept of a business object data guide structure. Such data guide helps to store the information which sub-structures of a business object are populated with data. A possible implementation of such an object data guide is a simple bit mask. Our example in Fig. 34.1, contains such a bit mask at the root node of the *sales order* business object. The zero at the second position of the object data guide indicates that the *delivery terms* sub-structure is not filled with data, thus joining with that table can be omitted.

34.1 Object-Relational Mapping

Object-orientation and relational algebra are based on different concepts. Object-orientation uses a navigational model that allows to traverse the object graph by following associations. Relational algebra uses set operations to declaratively determine specific relations. Based on these fundamental differences, a number of problems arises, which are commonly referred to as the *object-relational impedance mismatch* [CI05]. One approach to bridge the object-relational gap is Object-Relational Mapping (ORM). ORM is the process of defining a set of transformation rules that map the structural aspects (i.e. fields and associations) of an object to a set of database tables and vice versa [Rus08]. Using an ORM can help to reduce boilerplate code, because it automatically generates mapping code, based on the defined transformation rules, and it provides an object-oriented interface to a relational database. Although developers find the idea of interacting with a relational database in an object-oriented fashion tempting, it results in a number of difficulties. The power of set-oriented data processing with SQL has a high importance in the context of business functionality. Especially aggregations and joins provide efficient means to express calculations common to many facets of an enterprises' core functions. Relying solely on the object-oriented navigational model an ORM provides to access the data may result in data access patterns with performance issues compared to optimized SQL. Ted Neward elaborates on some of the potential pitfalls of introducing an ORM to a system's architecture in [New06]. Early success, underestimated data model complexity, and constant data model modifications reveal the danger of an absolute commitment to the use of an ORM. The use of an ORM allows to reduce complexity and increases maintainability of a software system. However, its application should be considered carefully in the context of the design of a data intensive software system.

34.2 Self Test Questions

1. **Object-Orientation**
   How does OO help in designing enterprise applications?

   (a) OO concepts, such as encapsulation, aggregation, and inheritance help to design domain model, which can be discussed and validated with domain experts.
   (b) OO allows domain experts to validate the data model of the application.
   (c) OO provides the highest performance with regards to mathematical algorithms.
   (d) OO provides a seamless integration of declarative concepts e.g., SQL.

References

[CI05]   W.R. Cook, A.H. Ibrahim, Integrating programming languages and databases: What is the problem. *In ODBMS.ORG, Expert Article* (2005)

[DN66]   O.-J. Dahl, K. Nygaard, Simula: An algol-based simulation language. Comm. ACM **9**(9), 671–678 (1966)

[Ing78]   D.H.H. Ingalls, The smalltalk-76 programming system design and implementation, in *Proceedings of the 5th ACM SIGACT-SIGPLAN Symposium on Principles of Programming Languages*, POPL '78 (ACM, New York, 1978), pp. 9–16

[New06]   T. Neward, The Vietnam of computer science, **6** (2006). http://blogs.tedneward.com/2006/06/26/The+Vietnam+Of+Computer+Science.aspx

[Rus08]   C. Russell, Bridging the object-relational divide. Queue **6**(3), 18–28 (2008)

Chapter 35
Examples for New Enterprise Applications

This chapter gives an overview of exemplary enterprise applications that build on the changed foundations and database techniques described in this book. We will first describe how in-memory technology can be leveraged for the interactive analysis of point-of-sales data, followed by an example application for high performance in-memory genome analyses.

35.1 Interactive Analyses of Point-of-Sales Data

Retailers face not only the challenge of consolidating all the data generated by electronic point-of-sale (POS) terminals, but also to leverage the data to derive business value. Especially when the data is stored at its finest granularity recording the actual transactions with all their items, processing becomes a challenge. In a prototypical implementation described in [SFKP14], we show how in-memory technology can help to analyze POS data and how it enables new types of enterprise applications. We show that it is possible to interactively explore the transactional data set without precomputing analytical summaries while providing users with full flexibility. We present a prototypical application for interactive analyses and exploration of 8 billion records of real data from a large retail company with subsecond response times.

35.1.1 Motivation

Information derived from POS data can be used for the planning and verification of sales promotions, cross-selling, up-selling, loyalty programs or product placement inside stores.

As an example, we will focus on the planning of sales promotions. Consumer sales promotions are an essential part of a retailers business, as the frequency of promotions increased significantly over the last years through competition and price pressure. Nowadays, up to 20% of the total revenue of German retailers are generated through promotions [SAP11]. When planning promotions, extended analyses based on the POS data are taken into account.

However, the amount of data and the separation of today's analytical and transactional systems make it difficult to perform these analyses and limit the range of questions that can be answered with the analytical system. In the following, we outline the advantages of in-memory technology and how it can be used for interactive analyses of the transactional POS data.

35.1.2 POS Explorer

The speed and flexibility provided by the technical foundation of in-memory database technology enables the creation of completely new tools for enterprises. To illustrate the potentials of in-memory technology, we built the POS Explorer application for the interactive analysis of POS data as a proof of concept to outline the potentials of such applications.

The application is designed to support the planning and verification of promotions, while pursuing the following goals. The application should (a) work directly on the POS transactions as the finest level of granularity, (b) enable interactive access to the data with sub-second response times, (c) provide full flexibility to users and not limit queries due to missing pre-aggregated information and (d) leverage a modern javascript and HTML based user-interface for access in browsers and mobile devices.

The main functionality provided by the application is (A) a hierarchical product browser with live calculated key-figures and full filter flexibility, (B) an interactive basket analysis for selected products and (C) an interactive heat-map analysis of sales and revenue.

Transactional Data Model and Real World Data Volume

The paradigm of analyzing the transactional data as it is recorded on the level of single transactions is a fundamental design decision and in stark contrast to current approaches [TLNJ11]. Typically, enterprises use IT-landscapes with a variety of systems and replicate data between those systems while transforming data and pre-calculating aggregated information [BFH+11]. The results are then used in analytical systems or for the generation of reports summarizing the information. However, it is expensive, complicated and slow to establish additional reports if the ETL process has to be adapted.

Therefore, all operations in our demo application work directly on the transactional data model, which stores single POS terminal transactions. The customer transactions are stored as a set of items, which are connected through unique transaction identifiers. Every item contains information about the specified product, the price, the sold quantity, the total revenue and if the product was sold during a promotion. Each receipt has additional information, such as the store, date and time of the transaction.

The application was developed based on a real data set from a large retail company, containing 8 billion line items from sales over 2 years, 500,000 products and 5,000 stores. The uncompressed size of the dataset is roughly one terabyte. As a back of the envelope calculation and assuming a hard disk bandwidth of 100 megabytes per second per disk, one would need about 10,000 disks to scan the whole dataset in a traditional disk-based row store in one second. However, the prototype application runs on a cluster of 25 nodes with 1,000 processor cores and 25 terabytes of main memory.[1] The processing power and level of parallelization of the cluster correspond to the hardware that we expect to be generally available on a single, affordable high-end computer in a few years.

On-the-Fly Calculation of Key Figures

As an example how the flexibility provided by in-memory technology can be leveraged, the demo application allows users to browse through products while calculating common key figures on the fly. Figure 35.1 shows a screenshot from the demo application's hierarchical product browser with the key figures relative market share, generated revenue and the basket value as well as their development over time.

The relative market share indicates the percentage of a product's sales within its product group. It is calculated by summing up all sales of products in a certain product group and comparing it with the sales of each product within that group to calculate the relative share. The generated revenue is calculated by the sum of all sales of the respective product. The basket value indicates the average value of transactions that contain a product and is calculated by the average basket size and basket value. This allows users to quickly identify products or product groups that drive revenue as the basket value is often a good indicator for products to promote.

While the user navigates through the product hierarchy or directly searches for a product, every click results in multiple queries scanning the complete data set and calculating the requested numbers. As everything is always calculated directly on the POS data and no aggregates are used, the data can be filtered on any attribute or combination of attributes, for example the size of stores, periods of time, product groups or areas. In the application, these filters can be adjusted directly by the user through the optional menu at the top, providing a fixed set of filters. However, it would also be possible to directly expose the complete data schema to users, allowing them to filter on any available attribute.

---

[1] Hosted at the Future SOC Lab at the Hasso Plattner Institute.
http://www.hpi-web.de/future_soc_lab.

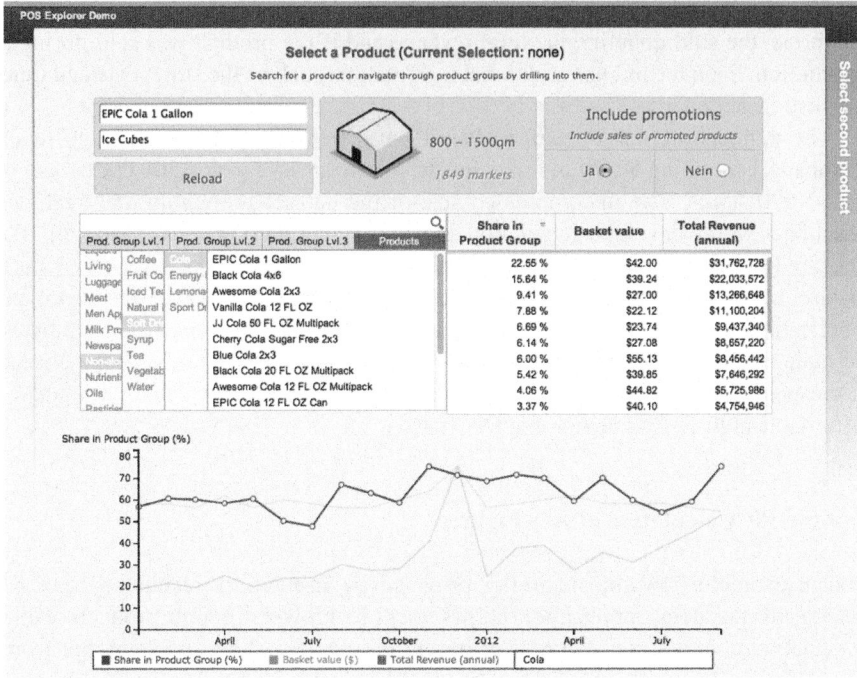

Fig. 35.1 The prototypical application enables an interactive exploration of the transactional POS data set with 8 billion records. The screenshot shows the entry point of the tool, presenting a hierarchical product browser with key figures for every product and their development over time. Each selection step within the product hierarchy aggregates all transactions and is done interactively with response times below 1 second.

Despite the huge amount of data and without using pre-aggregated summaries, every query is processed with sub-second response times. The application allows to browse the presented products interactively and users can enter into a dialog with the system to quickly retrieve answers and to articulate follow up questions. This enables the exploration of data in a try and error fashion, assisting in the discovery of new insights.

Heat-Map Analysis

The calculation of key figures on the actual transactions allows for new types of analyses that take the additional information into account. As an example, we implemented a heat map visualization over the days of the week and the hours of the day, as shown in Fig. 35.2. This particular visualization is useful for the verification of promotions. It is an example of data analyses that are only feasible with fast access to the raw POS data. We can visualize hourly aggregates for the common sales of any two products, while allowing users to interactively filter the considered

Fig. 35.2 Intuitive visualization of a heat map analysis of a product's revenue over the days and hours of the week. The analysis leverages the granularity of data at the transaction level and provides interactivity through sub-second query processing even on 8 billion records.

time-range, store-size and region without the need for pre-aggregated cubes that would lead to immense storage and maintenance costs.

Basket Analysis

A basket analysis determines relationships of products that are most commonly sold together. Instead of using old traditional list views, Fig. 35.3 shows a new intuitive visualization of the result of a basket analysis. The circle visualizes the relationships of the most commonly sold products with a product of interest, whereas the size of the area can be used to visualize another dimension as for example the average basket size or value. Like the calculation of key figures, the basket analysis runs interactively and can be executed on any arbitrarily filtered data set. The quality of a rule $A \rightarrow B$ can be drawn from its *support* and *confidence* values. The support value refers to the number of transactions that contain product A, whereas the confidence value reflects the number of transactions that contain product A and B.

Since the receipts are distributed across the computing cluster based on the transaction identifier, the database system can evaluate the intermediate results on each partition and aggregate only the counts, thereby keeping the network traffic between the computing-nodes low and resulting in low response times.

Advantages over Traditional Approaches

In traditional analytical systems the set of possible questions and their detail level are defined at the implementation and execution of the Extract-Transform-Load (ETL) process. The trade-off is between a level of detail and storage and processing

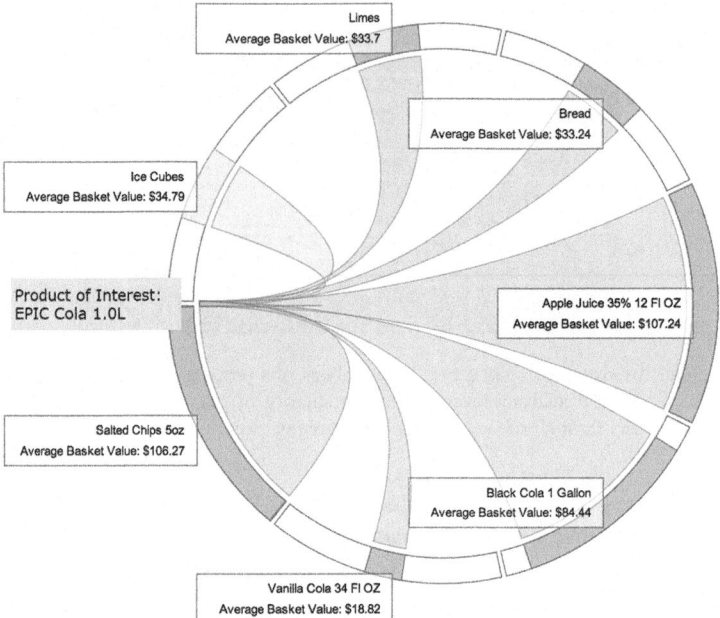

Fig. 35.3 Intuitive visualization of the result of a basket analysis, whereas the *circle* visualizes products that are most commonly sold together with the product of interest and the size of the area indicates the average basket values.

requirements for creating the data. As a result, the cubes contain the data up to a certain detail level in all dimensions. With a product hierarchy of 500,000 products, any cube that cross-relates all of these products would contain 250 billion data fields for each dimension. In this setting, storing a relation between two products on a monthly level would require 3 trillion data fields for 1 year. Due to these effects, today's basket analyses have to focus on subsets of products, e.g. to analyze the products of a few product groups at a time.

The main point of our solution is not that the presented analyses were not possible before, but that users are now able to perform these analyses interactively with sub-second response times even on large amounts of data and that the user has full flexibility and can filter and aggregate data on the fly as needed.

35.2   High Performance In-Memory Genome Data Analysis

Precision medicine aims at treating patients specifically based on individual dispositions, e.g., genetic or environmental factors [Jai09]. For that, researchers and physicians require a holistic view on all relevant patient specifics when making treatment decisions. Thus, the detailed acquisition of medical data is the foundation

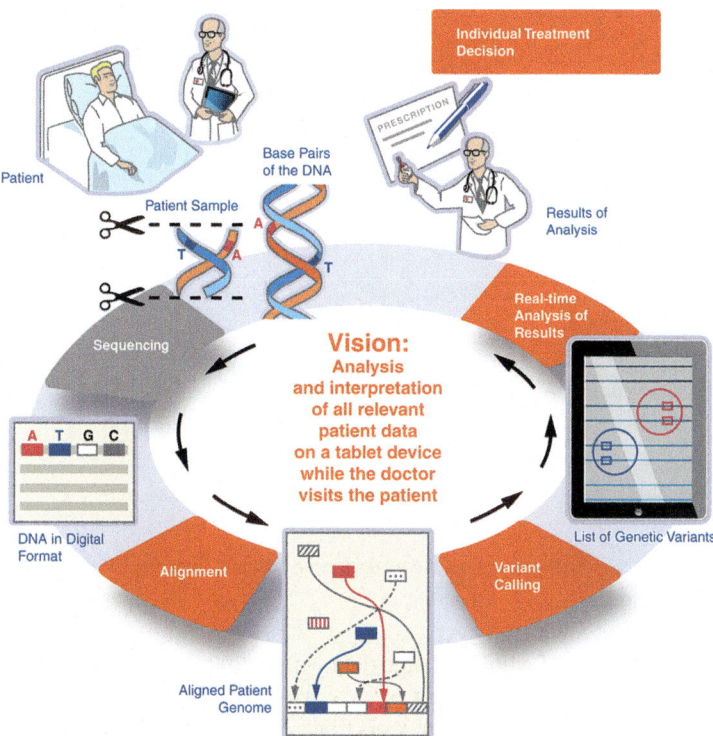

Fig. 35.4 Data processing steps involved in the analysis of genome data. Sequencing the samples results in chunks of DNA available in digital form. During alignment their position within the whole genome is mapped. Variant calling results in a list of differences compared to a fixed reference. The analysis obtains new insights based on the list of detected variants.

for personalized therapy decisions. The more fine-grained the available data is, the more specific the gained insights will be, but the complexity of data processing will rise as well. This requires tool support to identify the relevant portion of data out of the increasing amount of acquired diagnostic data [PS14].

Figure 35.4 depicts the genome data workflow in the course of precision medicine. After a sample has been acquired, it is sequenced, which results in short chunks of Deoxyribonucleic Acid (DNA) in digital form. The DNA chunks need to be aligned to reconstruct the whole genome and variants compared to a reference, e.g., normal vs. pathologic tissue, are detected during variant calling. The analysis of genome data builds on the list of detected variants, e.g., to identify driver mutations for a medical finding [B+10].

Figure 35.5 provides a comparison of costs for sequencing and main memory modules on a logarithmic scale. Both charts follow a steadily declining trend, which facilitates the increasing use of Next-Generation Sequencing (NGS) for whole genome sequencing and IMDB technology for data analysis. Latest NGS devices

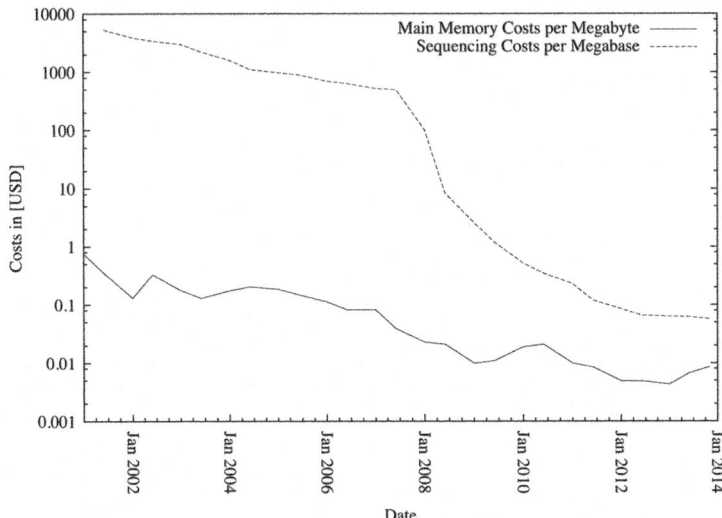

Fig. 35.5 Costs for next-generation sequencing and main memory 2001–2014 adapted from [Nat13, McC13]

enable the processing of whole genome data within hours at reduced costs [Ans09]. The time consumed for sequencing is meanwhile a comparable small portion of the time consumed by the complete workflow. As a result, data processing and its analysis consume a significantly higher portion of the time and accelerating them will affect the overall workflow duration.

We focus on how to optimize the time-consuming data processing and analysis aspects of the workflow by combining latest software and hardware trends to create an integrated software system, which supports life science experts in their daily work.

Our High Performance In-memory Genome (HIG) system architecture modeled as Fundamental Modeling Concept (FMC) block diagram is depicted in Fig. 35.6 [KGT06]. It combines data from various data sources, such as patient-specific data, genome data, and annotation data, within a single system to enable flexible real-time analysis and combination. In the following, the system layers are described in further detail.

35.2.1   Application Layer

The application layer consists of special purpose applications to answer medical and research questions [PS14].[2] We provide an Application Programming Interface

---

[2]You can access our cloud services online at http://www.analyzegenomes.com.

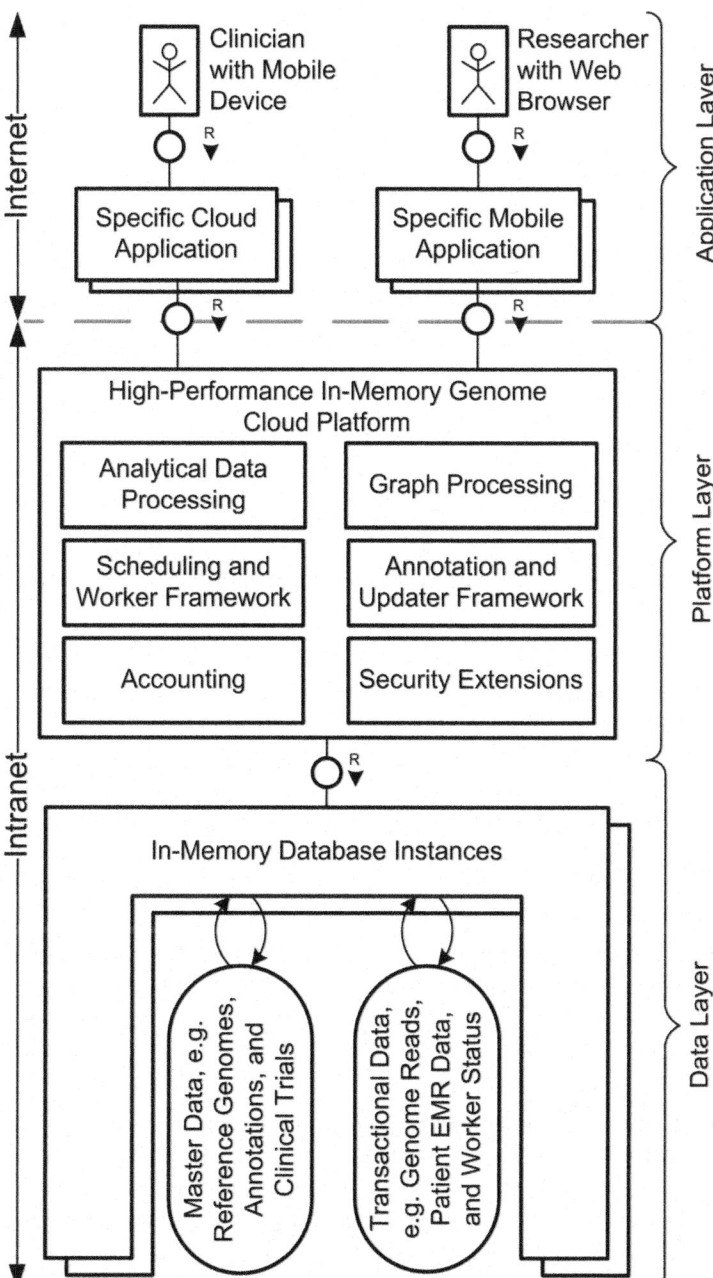

Fig. 35.6 Our system architecture consist of application, platform, and data layer. Analysis and processing of data is performed in the platform layer, which eliminates time-consuming data transfer

Fig. 35.7 The patient-specific clinical trials search results based on the individual anamnesis of a patient. Relevant entities from the free-text description of a clinical trial are extracted with the help of specific text-mining rules

(API) that can be consumed by various kinds of applications, such as web browser or mobile applications. Figure 35.6 depicts the data exchange via asynchronous Ajax calls and JavaScript Object Notation (JSON) [Hol08, Cro06]. As a result, performing specific analyses is no longer limited to a specific location, e.g., the desktop computer of a clinician. Instead, all applications can be accessed via devices connected to the Internet, e.g., laptop, mobile phone, or tablet computer. Thus, having access to relevant data at any time enhances the user's productivity. The end user can access these cloud applications via any Internet browser after registration. Selected cloud applications are our cohort analysis and clinical trials search, which are described in further detail in Sects. 35.2.1 and 35.2.1 respectively.

Clinical Trials Application

Our clinical trials search assists physicians in finding adequate clinical trials for their patients. It analyses patient data, such as age, gender, preconditions and existing mutations, and matches them with clinical trials descriptions from clinicaltrials.gov [U.S13a]. Our analysis incorporates more than 130,000 clinical trial descriptions, which are processed and ranked in real-time accordingly to the personal anamnesis of each individual patient. The ranked results are summarized on a single screen and provided to the researcher as depicted in Fig. 35.7.

The clinical trials search incorporates the extraction of entities and features from the textual description. For that, we assembled a set of customized vocabulary. For example, we use a vocabulary set for human gene identifiers with more than 120,000 gene names and synonyms and a vocabulary set for pharmaceutical ingredients

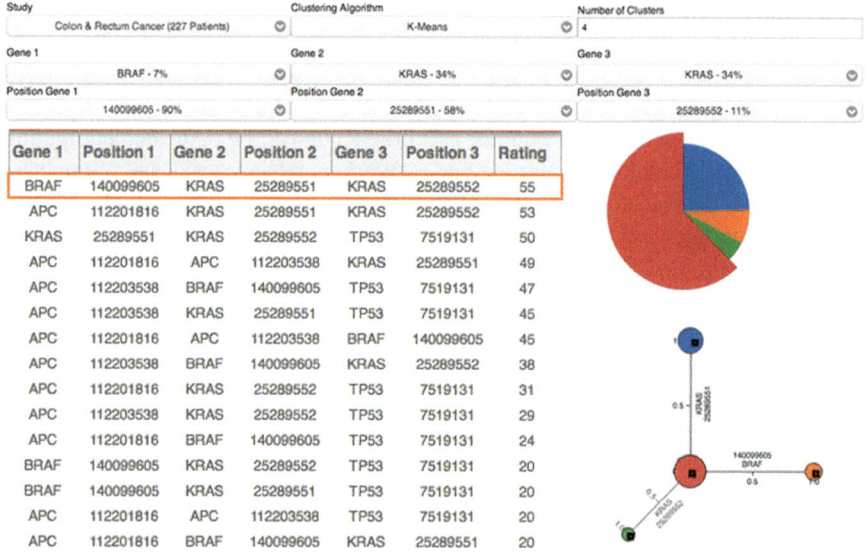

Study			Clustering Algorithm			Number of Clusters		
Colon & Rectum Cancer (227 Patients)		⊘	K-Means		⊘	4		
Gene 1			**Gene 2**			**Gene 3**		
BRAF - 7%		⊘	KRAS - 34%		⊘	KRAS - 34%		⊘
Position Gene 1			**Position Gene 2**			**Position Gene 3**		
140099605 - 90%		⊘	25289551 - 58%		⊘	25289552 - 11%		⊘

Gene 1	Position 1	Gene 2	Position 2	Gene 3	Position 3	Rating
BRAF	140099605	KRAS	25289551	KRAS	25289552	55
APC	112201816	KRAS	25289551	KRAS	25289552	53
KRAS	25289551	KRAS	25289552	TP53	7519131	50
APC	112201816	APC	112203538	KRAS	25289551	49
APC	112203538	BRAF	140099605	TP53	7519131	47
APC	112203538	KRAS	25289551	TP53	7519131	45
APC	112201816	APC	112203538	BRAF	140099605	45
APC	112203538	BRAF	140099605	KRAS	25289552	38
APC	112201816	KRAS	25289552	TP53	7519131	31
APC	112203538	KRAS	25289552	TP53	7519131	29
APC	112201816	BRAF	140099605	TP53	7519131	24
BRAF	140099605	KRAS	25289552	TP53	7519131	20
BRAF	140099605	KRAS	25289551	TP53	7519131	20
APC	112201816	APC	112203538	TP53	7519131	20
APC	112201816	BRAF	140099605	KRAS	25289551	20

Fig. 35.8 The results of an interactive analysis of a cohort of colon carcinoma patients using k-means clustering. It proposes relevant combinations of genomic loci, such as gene KRAS on chromosome 12 at position 25,289,551, which are present in the majority of cohort members as depicted by the pie chart on the right

with more than 7,000 pharmaceutical ingredients. Our ontologies incorporate a set of standardized vocabularies, e.g., the Metathesaurus Structured Product Labels (MTHSPL) of the Unified Medical Language System (UMLS) [U.S13b].

Patient Cohort Analysis

Our cohort analysis application enables researchers and clinicians to perform interactive clustering on the data stored in the IMDB, e.g., k-means and hierarchical clustering [Kra09, Chap. 13]. Thus, researchers and clinicians are able to verify hypotheses by combining patient and genome data in real-time. Therefore, they use patient-specific and genome data loaded into the IMDB and perform the interactive cohort analysis as part of the platform layer as depicted in Fig. 35.8.

35.2.2   Platform Layer

The platform layer holds the complete process logic and consists of the IMDB system for enabling real-time analysis of genomic data. We developed specific extensions for the IMDB system to support genome data processing and its analysis. In the following, selected extensions and their integration in the IMDB system are described in more detail.

Scheduling of Data Processing

We extended the IMDB by a worker framework, which executes tasks asyn-chronously, e.g., alignment of chunks of genome data. The framework consists of a task scheduler instance and a number of workers controlling dedicated computing resources, e.g., individual computing nodes. Workers retrieve tasks and parameters by the scheduler instance and perform specific tasks, such as workbench prepara-tion, task execution, and maintenance of status information. Thus, all workers are connected to the IMDB to store status information about currently executed tasks.

Furthermore, the scheduler supervises the responsiveness of individual compute resources. If a predefined response behavior is no longer guaranteed, e.g., due to an overloaded compute node or a crashed worker process, workers are marked as unresponsive. As a result, work-in-progress tasks of the unresponsive worker are reassigned to a new worker and this worker is scheduled for a restart.

Updater Framework

We consider the use of latest international research results as enabler for evidence-based therapy decisions [SHP13]. The updater framework is the basis for combining international research results. It periodically checks all registered Internet sources, such as public FTP servers or web sites, for updated and newly added versions of annotations, e.g., database exports as dumps or characteristic file formats, such as CSV, TSV, and VCF. If the online version is newer than the locally available version, the new data is automatically downloaded and imported in the IMDB to extend the knowledge base.

The import of new versions of research databases is performed as a background job without affecting the system's operation. We import new data without any data transformations in advance. Thus, it is available for real-time analysis without any delay [BSP12, FCP$^+$12]. For example, the following selected research databases are checked regularly by our updater framework: National Center for Biotechnology Information (NCBI), Sanger's catalogue of somatic mutations in cancer, University of California, Santa Cruz (UCSC) [Nat, F$^+$10, M$^+$12].

35.2.3 Data Layer

The data layer holds all required data for performing processing and analyzing of genomic data. The data can be distinguished in the two categories of master and transactional data [DM11]. For example, human reference genomes and annotation data are referred to as master data, whereas patient-specific NGS data and Electronic Medical Records (EMR) are referred to as transactional data [The, RNK91]. Its analysis is the basis for gathering specific insights, e.g., individual genetic dispositions, and for leveraging personalized treatment decision in the course of personalized medicine [Jai09].

35.2.4 Summary

Nowadays, a range of time-consuming tasks has to be accomplished before researchers and clinicians can work with analysis results, e.g., to gain new insights. In the following, selected tasks are outlined:

- Genome data of patients has to be analyzed in customized workflows, e.g., to identify genetic variants relevant for a certain disease,
- Researchers and clinicians have to gather and combine relevant information from distributed, heterogenous data sources and link them to patient data, and
- Researchers and clinicians need to apply gained insights on their patient data, e.g., to formulate new hypotheses and to verify them.

Nowadays, almost all aforementioned tasks have to be carried out manually. As a result, it takes a considerable amount of time until an optimal treatment decision can be made.

With our HIG platform, we redefine and accelerate the process of finding treatment decisions. We provide the infrastructure and computational resources to run configurable genome data analysis workflows. Furthermore, we have created a knowledge base by integrating content from various heterogenous data sources, which are updated constantly to always provide latest findings from research. With the capabilities of the applied in-memory technology, we can immediately combine tremendous amounts of data, taking also advantage of the whole spectrum of analysis features provided, e.g., for textual analysis. With our innovative approach, gaining new insights on patient data shifts from a time consuming, manual process to immediate, interactive data exploration.

As a result, researchers and clinicians can spent additional time on formulating hypotheses, their verification, and drawing corresponding conclusions to come up with more accurate treatment decisions.

References

[Ans09]   W.J. Ansorge, Next-generation DNA sequencing techniques. New Biotechnol. **25**(4), 195–203 (2009)

[B+10]   I. Bozic et al., Accumulation of driver and passenger mutations during tumor progression. Proc. Natl. Acad. Sci. USA **107**(43), 18545–18550 (2010)

[BFH+11]   B. Bläser, K. Fleckstein, S. Hoffmann, O. Schall, O. Rutz, SAP enterprise data warehouse for point of sales data optimized for IBM DB2 for Linux, UNIX, and Windows on IBM Power Systems. SAP SCN, DOC-14457 (2011)

[BSP12]   A. Bog, K. Sachs, H. Plattner, Interactive Performance Monitoring of a Composite
            OLTP and OLAP Workload, in *Proceedings of the Int'l Conf on Mgmt of Data 2012*
            (ACM, Scottsdale, AZ, 2012), pp. 645–648
[Cro06]   D. Crockford, RFC4627: The application/json Media Type for JavaScript Object
            Notation (JSON). http://www.ietf.org/rfc/rfc4627.txt [retrieved: Sep3, 2012], July
            2006
[DM11]   T.K. Das, M.R. Mishra, A study on challenges and opportunities in master data
            management. Int. J. Database Mgmt Syst. **3**(2), 129–139 (2011)
[F⁺10]   S.A. Forbes et al., The catalogue of somatic mutations in cancer: a resource to
            investigate acquired mutations in human cancer. Nucl. Acids Res. **38**, D652–D657
            (2010)
[FCP⁺12]   F. Färber, S.K. Cha, J. Primsch, C. Bornhövd, S. Sigg, W. Lehner, SAP HANA
            database: data management for modern business applications. SIGMOD Rec. **40**(4),
            45–51 (2012)
[Hol08]   A.T. Holdener, *AJAX: The Definitive Guide*,1st edn. (O'Reilly, California, 2008)
[Jai09]   K.K. Jain, *Textbook of Pers. Medicine* (Springer, New York, 2009)
[KGT06]   A. Knopfel, B. Grone, P. Tabeling, *Fundamental Modeling Concepts: Effective
            Communication of IT Systems* (Wiley, West Sussex, 2006)
[Kra09]   S.A. Krawetz, *Bioinformatics for Systems Biology* (Humana Press, New York, 2009)
[M⁺12]   L.R. Meyer et al., The UCSC genome browser database: extensions and updates
            2013. Nucl. Acids Res., D64–D69 (2012)
[McC13]   J.C. McCallum, Memory Prices (1957–2013). http://www.jcmit.com/memoryprice.
            htm[retrieved: Feb 11, 2014], Feb 2013
[Nat]   National Center for Biotechnology Information, All Resources. http://www.ncbi.nlm.
            nih.gov/guide/all/[retrieved: Jan 5, 2013]
[Nat13]   National Human Genome Research Institute, DNA Sequencing Costs. http://www.
            genome.gov/sequencingcosts/[retrieved: Feb 11, 2014], Apr 2013
[PS14]   H. Plattner, M.-P. Schapranow (eds.), *High-Performance In-Memory Genome Data
            Analysis: How In-Memory Database Technology Accelerates Personalized Medicine*
            (Springer, Heidelberg, 2014)
[RNK91]   A.L. Rector, W.A. Nolan, S. Kay, Foundations for an electronic medical record.
            Methods Inform. Med. 179–186 (1991)
[SAP11]   SAP AG und GfK Panel Services Deutschland, Professionalisierung des Promotions-
            managements im deutschen LEH. *Eine Gemeinschaftsstudie von GfK und SAP* (2011)
[SFKP14]   D. Schwalb, M. Faust, J. Krueger, H. Plattner, Leveraging in-memory technology for
            interactive analyses of point-of-sales data. *Workshop on Big Data Customer Analytics
            in Conjunction with ICDE*, 2014
[SHP13]   M.-P. Schapranow, F. Häger, H. Plattner, High-performance in-memory genome
            project: a platform for integrated real-time genome data analysis, in *Proceedings
            of the 2nd International Conference on Global Health Challenges* (IARIA, 2013),
            pp. 5–10
[The]   The Genome Reference Consortium, Genome Assemblies. http://www.ncbi.nlm.nih.
            gov/projects/genome/assembly/grc/data.shtml[retrieved: Jan 5, 2013]
[TLNJ11]   T.J. Teorey, S.S. Lightstone, T. Nadeau, H.V. Jagadish, *Database Modeling and
            Design: Logical Design*, 5th edn. The Morgan Kaufmann Series in Data Management
            Systems (Elsevier Science, Burlington, 2011)
[U.S13a]   U.S. National Institutes of Health, Clinicaltrials.gov. http://www.clinicaltrials.
            gov/[retrieved: Jul 17, 2013], 2013
[U.S13b]   U.S. National Library of Medicine, Unified Medical Language System (UMLS).
            http://www.nlm.nih.gov/research/umls/[retrieved: Sep 19, 2013], Jul 2013

Chapter 36
Bypass Solution

As illustrated throughout the course, in-memory data management can enable significant advantages to data processing within enterprises. However, the transition for enterprise applications to an in-memory database will require radical changes to data organization and processing, resulting in major adaptations throughout the entire stack of enterprise applications. By considering conservative upgrade policies used by many ERP system customers, the adoption of in-memory technology is often delayed, because such radical changes do not align well with the evolutionary modification schemes of business-critical customer systems. Consequently, a risk-free approach is required to help enterprises to immediately leverage in-memory data management technology without disruption of their existing enterprise systems.

We propose a transition process that allows customers to benefit from in-memory technology without changing their running systems. This is a step by step, non-disruptive process that helps to transform traditionally separated operational and analytical systems into what we believe is the future for enterprise applications: transactional and analytical workloads handled by a single, in-memory database.

Within the first step of the transition, an in-memory database will run in parallel to the traditional database and the data will be stored in both systems. Secondly, new side-by-side applications using transactional data of the ERP system can be developed. In the next step, a new Business Intelligence (BI) solution can be introduced that is able to answer flexible, ad-hoc queries and operate without materialized views, aggregating all necessary information on the fly from the transactional data. Complex ETL processes are not required any longer. Finally, the traditional disk-based database of the ERP system can be replaced by a column-oriented dictionary-encoded in-memory database. This transition is described in more detail in the next section.

H. Plattner, *A Course in In-Memory Data Management*,
DOI 10.1007/978-3-642-55270-0_36, © Springer-Verlag Berlin Heidelberg 2014

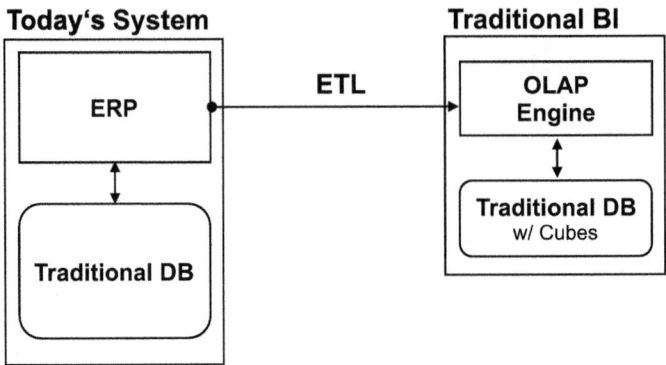

Fig. 36.1 Initial architecture

36.1 Transition Steps in Detail

First of all, we start with a commonly found initial architecture of an existing enterprise solution as illustrated in Fig. 36.1.

Typically, it consists of multiple OLTP and OLAP systems, each of them running on separate databases. The OLAP system consolidates data from multiple OLTP systems and external data sources. A costly and time-consuming ETL process between the OLTP and OLAP systems is used to pre-aggregate data for the OLAP system. Based on this architecture, the non-disruptive transition plan that we call "bypass solution" has been developed.

In the first step of this approach (see Fig. 36.2), the IMDB is installed and connected to the traditional database.

The only difference will be in data representation: data will be stored in columns. An initial load to the in-memory database (IMDB) creates a copy of the existing system state with all business objects in the IMDB. In spite of the huge volume of data to be reproduced, first experiments with massively parallel bulk loads of customer data have shown that even for the largest companies this one-time initialization can be done in only a few hours.

After the initial load, the two storages will be maintained in parallel, every document and change in a business object is stored in both databases. For this, established database replication technologies are used. The high compression rate in a column store helps to decrease the amount of main memory required for such parallel use of two databases, and hence it does not lead to a significant waste of resources. At the same time, using the parallel installation of the IMDB, we can estimate performance and memory consumption benefits of this architecture for concrete business cases and prove the need of moving the system to the new data storage.

In a second step, new applications can be developed leveraging the potentials of the new technology (see Fig. 36.3). These applications only read the replicated ERP

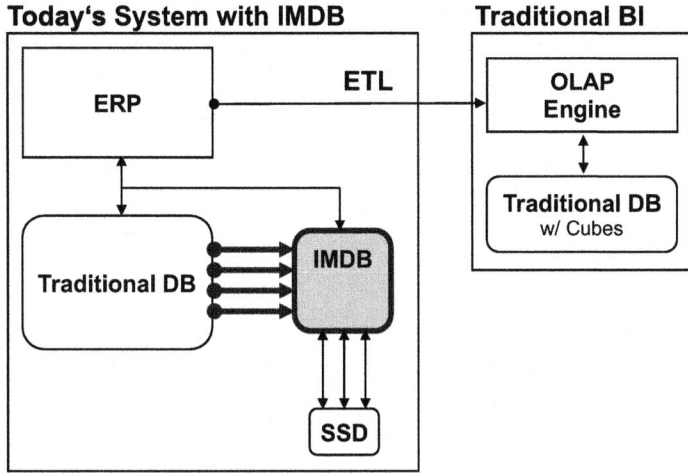

Fig. 36.2 Run data replication to the IMDB in parallel

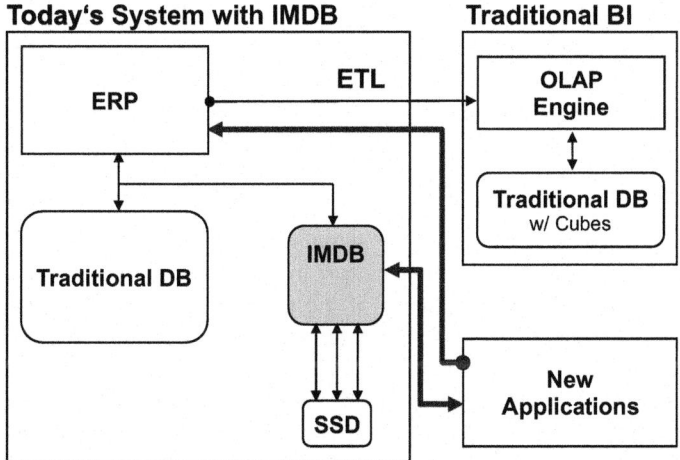

Fig. 36.3 Deploy new applications

system data. If they want to write data, they either use the existing interfaces of the ERP system or, if they want to store additional data, they use a separate segment in the in-memory database. That way, business value by new revolutionary applications can be generated from the first day on.

In a third step, which can be done in parallel with the previous ones, the BI system is ported to the IMDB as illustrated in Fig. 36.4. This will help to achieve another gain in reporting performance in comparison to disk-based OLAP systems.

The difference to the traditional system is that all materialized data cubes and aggregates will be removed. Instead, aggregates are computed on the fly and all

Today's System with IMDB   **Traditional BI with IMDB**

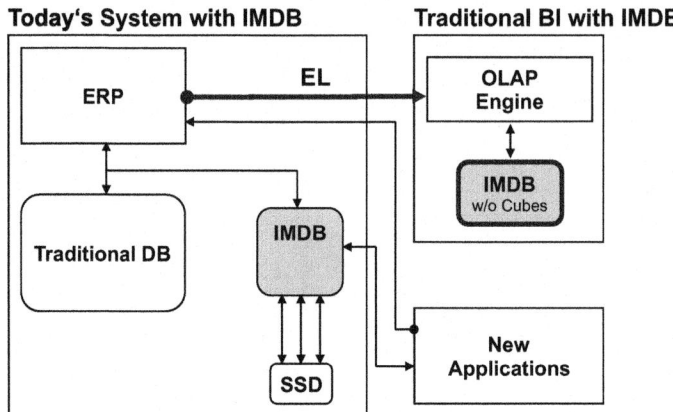

Fig. 36.4 Traditional BI system on IMDB

data-intensive operations are pushed to the database level. In comparison with storing all materialized aggregates and indices, this reduces the amount of main memory, which is necessary for the OLAP system. It immediately leads to the following benefits: the data cubes in a traditional BI system are usually updated on a weekly basis or even less frequently. However, executives, management, and all other decision makers often demand up-to-date information. By running OLTP and OLAP systems on the same IMDB platform, this information can be delivered in real time. Another advantage is that the ETL process is radically simplified as the complex calculation of aggregates is omitted. The ETL process can partly be replaced by the same replication mechanism used between the traditional ERP database and the parallel installation of the in-memory database. Therefore, we called the replication in Fig. 36.4 simply "EL", since no transformation takes place any longer, just extraction and load remain. Furthermore, the Business Intelligence (BI)-system will be more flexible as complex cube management and maintenance operations are abandoned and complexity is reduced.

In most cases, the migration to the in-memory database BI system can be conducted automatically. This is relatively simple as existing materialized views are replaced by non-materialized views. Analytical queries can be rewritten by generators. However, in complex migration scenarios, factors such as different SQL dialects, data structures, or software parts that rely on additional data in proprietary formats may pose an obstacle and might require manual interventions.

The final step of the suggested solution can be executed when the customer is comfortable with the parallel in-memory solution. In this step, the traditional OLTP database is switched off. After that, the customer works only with the consolidated in-memory enterprise system that is used for both transactional and analytical queries (see Fig. 36.5).

The BI system could theoretically be replaced in a setup with only one ERP system. Reality shows that many OLTP systems exist and external data has to be

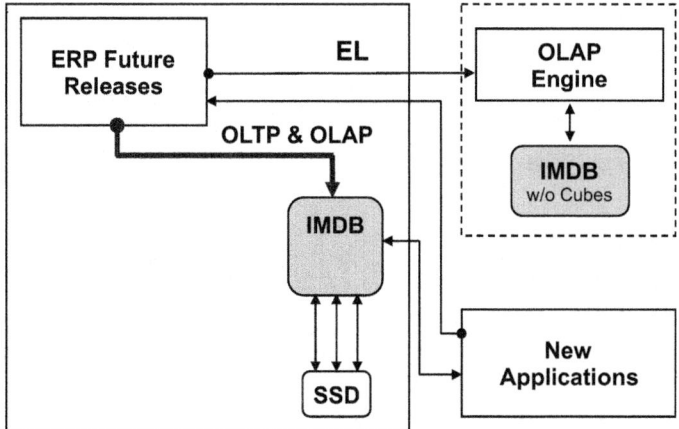

Fig. 36.5 Run OLTP and OLAP on IMDB

integrated into the analytical system. To this end, the traditional BI system is used as a data platform consolidating data from these OLTP systems and external sources.

Additionally, during system evolution, new extensions to the data model are possible. Adding new tables and new attributes to existing tables in the column store is done on the fly and this speeds up release cycles significantly.

36.2 Bypass Solution: Conclusion

As discussed above, the suggested bypass solution introduces a risk-free and non-disruptive transition to the in-memory database technology. The importance of this transition can be backed-up by selected customer examples: For a large financial service provider, the analysis of 33 million customer records could be reduced from 45 min on the traditional DBMS to 5 s on an IMDB. This increase in speed fundamentally changes the company's opportunities for customer relationship management, promotion planning, and cross selling.

In a similar use case, a large vendor in the construction industry is using an IMDB to analyze its nine million customer records and to create contact listings for specific regions, sales organizations, and branches. Customer contact listing is currently an IT process that may take 2–3 days to complete. A request must be sent to the IT department who must plan a background job that may take 30 min and the results have to be manually returned to the requester. With an IMDB, sales people can directly query the live system and create customer listings in any format they wish, in less than 10 s.

Concluding, the use of in memory technology can lead to a qualitative change in the business processes of an enterprise. The transaction that took days using the traditional process, can now be performed in the foreground on the fly. That will change the way of thinking and optimize many business processes in enterprises.

36.3   Self Test Questions

1. Transition to IMDBs

What does the transition to in-memory database technology mean for enterprise applications?

(a) The data organization will not change at all, but the source code of the applications has to be adapted
(b) Data organization and processing will change radically and enterprise applications need to be adapted
(c) There will be no impact on enterprise applications
(d) All enterprise applications are significantly sped up without incurring any adaptions

Self Test Solutions

Introduction

1. Rely on Disks
Does an in-memory database still rely on disks?
Possible Answers:

(a) Yes, because disk is faster than main memory when doing complex calculations
(b) No, data is kept in main memory only
(c) Yes, because some operations can only be performed on disk
(d) Yes, for archiving, backup, and recovery

Correct Answer: (d)
Explanation: Logs for archiving have to be stored on a persistent storage medium that preserves the content longer timespans. Main memory looses all information if the system is unpowered, therefore an other, persistent storage medium such as hard drives or SSDs has to be used for recovery and archiving.

New Requirements for Enterprise Computing

1. Data Explosion
Consider the formula 1 race car tracking example, with each race car having 512 sensors, each sensor records 32 events per second whereby each event is 64 byte in size.
How much data is produced by a F1 team, if a team has two cars in the race and the race takes 2 h?
Please use the following unit conversions: 1,000 byte = 1 kB, 1,000 kB = 1 MB, 1,000 MB = 1 GB.

Possible Answers:

(a) 14 GB
(b) 15.1 GB
(c) 32 GB
(d) 7.7 GB

Correct Answer: (b)
Explanation: Total time: $2\,h = 2 \cdot 60 \cdot 60\,s = 7{,}200\,s$
Total events per car: $7{,}200\,s \cdot 512$ sensors $\cdot 32$ events/second/sensor $= 117{,}964{,}800$ events
Total events per team: $(2 \cdot$ total events per car$) = 235{,}929{,}600$ events
Total amount of data per team: 64 byte/event $\cdot 235{,}929{,}600$ events $= 15{,}099{,}494{,}400$ byte ≈ 15.1 GB

Enterprise Application Characteristics

1. OLTP OLAP Separation Reasons
Why was OLAP separated from OLTP?
Possible Answers:

(a) Due to performance problems
(b) For archiving reasons; OLAP is more suitable for tape-archiving
(c) Out of security concerns
(d) Because some customers only wanted either OLTP or OLAP and did not want to pay for both

Correct Answer: (a)
Explanation: The runtimes of analytical queries are significantly higher than these of transactional ones. Based on this characteristic, analytical processing negatively affected the day-to-day business i.e. in terms of delayed sales processing. The separation of analytical and transactional queries to different machines was the inevitable consequence of the hardware and database prerequisites of these times.

Changes in Hardware

1. Speed per Core
What is the speed of a single core when processing a simple scan operation (under optimal conditions)?
Possible Answers:

(a) 4 GB/ms/core
(b) 4 MB/ms/core

(c) 4 MB/s/core

(d) 400 MB/s/core

Correct Answer: (b)

Explanation: Since a scan accesses memory sequentially, with the help of prefetching, cache misses can be avoided. Therefore we can assume that all data we have to deal with is in the Level 1 cache and ignore further delays incurred by fetching the data into the Level 1 cache. From Level 1 cache we need roughly 0.25 ns to load one byte, so in 1 ms, we can scan about 4,000,000 byte, which is 4 MB (1,000,000 ns / 0.25 ns/byte = 4,000,000 byte).

2. **Latency of Hard Disk and Main Memory**

Which statement concerning latency is wrong?

Possible Answers:

(a) The latency of main memory is about 100 ns

(b) A disk seek takes an average of 0.5 ms

(c) Accessing main memory is about 100,000 times faster than a disk seek

(d) Ten milliseconds is a good estimation for a disk seek

Correct Answer: (b)

Explanation: Please have a look at Table 4.1 on page 28.

A Blueprint for SanssouciDB

1. **New Bottleneck**

What is the new bottleneck of SanssouciDB that data access has to be optimized for?

Possible Answers:

(a) Disk

(b) The ETL process

(c) Main memory

(d) CPU

Correct Answer: (c)

Explanation: Main memory access is the new bottleneck, since the CPU busses limit overall data transfer. CPU speed increased according to Moore's Law (in terms of parallelism) and outperforms the bus speed. Disks are only used for backup and archiving reasons in SanssouciDB and are therefore not of interest for actual production usage. ETL processes are not of concern either, since all queries run on transactional data in one system for online transaction processing and online analytical processing.

2. Indexes

Can indexes still be used in SanssouciDB?

Possible Answers:

(a) No, because every column can be used as an index
(b) Yes, they can still be used to increase performance
(c) Yes, but only because data is compressed
(d) No, they are not even possible in columnar databases

Correct Answer: (b)

Explanation: Indices are a valid optimization of the performance in SanssouciDB. The index concept does not rely on compression.

Dictionary Encoding

1. Lossless Compression

For a column with few distinct values, how can dictionary encoding significantly reduce the required amount of memory without any loss of information?

Possible Answers:

(a) By mapping values to integers using the smallest number of bits possible to represent the given number of distinct values
(b) By converting everything into full text values. This allows for better compression techniques, because all values share the same data format.
(c) By saving only every second value
(d) By saving consecutive occurrences of the same value only once

Correct Answer: (a)

Explanation: The correct answer describes the main principle of dictionary encoding, which automatically results in a lossless compression if values appear more often than once. Saving only every second value is clearly lossy. The same applies for saving consecutive occurrences of the same value only once, if the quantity of occurrences is not saved as well. Additionally, this does not describe dictionary encoding, but Run-Length Encoding. Transforming numbers and other values into text values increases the data size, since each character value is at least 1 byte in order to allow the representation of the full alphabet. Number representations are usually limited to certain upper limits and achieve much smaller data sizes.

2. Compression Factor on Whole Table

Given a population table (50 millions rows) with the following columns:

- name (49 bytes, 20,000 distinct values)
- surname (49 bytes, 100,000 distinct values)

- age (1 byte, 128 distinct values)
- gender (1 byte, 2 distinct values)

What is the compression factor (uncompressed size/compressed size) when applying dictionary encoding?
Possible Answers:

(a) ≈ 20
(b) ≈ 90
(c) ≈ 10
(d) ≈ 5

Correct Answer: (a)
Explanation: Calculation without dictionary encoding:
Total size per row: $49 + 49 + 1 + 1$ (byte) $= 100$ byte
Total size: 100 byte $\cdot$ 50 million rows $= 5{,}000$ MB
Calculation with dictionary encoding:
Number of bit needed for the attributes:

- names: $log_2(20{,}000) < 15$
- surnames: $log_2(100{,}000) < 17$
- ages: $log_2(128) \leq 7$
- genders: $log_2(2) \leq 1$

Size of the attribute vectors:

- 50 million rows $\cdot$ $(15 + 17 + 7 + 1)$ bit $= 2{,}000$ million bit $= 250$ MB

Size of the dictionaries:

- names: $20{,}000 \cdot 49$ byte $= 980$ kB
- surnames: $100{,}000 \cdot 49$ byte $= 4.9$ MB
- ages: $128 \cdot 7$ byte $= 896$ byte
- genders: $2 \cdot 1$ byte $= 2$ byte

Total dictionary size: 4.9 MB $+ 980$ kB $+ 896$ byte $+ 2$ byte ≈ 5 MB
Overall size: size of attribute vectors $+$ size of dictionaries $= 250$ MB $+ 5$ MB $= 255$ MB
Compression rate:
$5{,}000$ MB $/ 255$ MB $= 19.6 \approx 20$

3. **Information in the Dictionary**
   What information is saved in a dictionary in the context of dictionary encoding?
   Possible Answers:

   (a) Cardinality of a value
   (b) All distinct values
   (c) Hash of a value of all distinct values
   (d) Size of a value in bytes

   Correct Answer: (b)

Explanation: The dictionary is used for encoding the values of a column. Therefore it consists of a list of all distinct values to be encoded and the resulting encoded values (in most cases ascending numbers). The distinct values are used to encode the attributes in user queries during look-ups and to decode the retrieved numbers from query results back to meaningful, human-readable values.

4. Advantages Through Dictionary Encoding

What is an advantage of dictionary encoding?

Possible Answers:

(a) Sequentially writing data to the database is sped up
(b) Aggregate functions are sped up
(c) Raw data transfer speed between application and database server is increased
(d) INSERT operations are simplified

Correct Answer: (b)

Explanation: Aggregate functions are sped up when using dictionary encoding. because less data has to be transferred from main memory to CPU. The raw data transfer speed between application and database server is not increased, this is a determined by the physical hardware and exploited to the maximum. Insert operations suffer from dictionary encoding, because new values that are not yet present in the dictionary, have to be added to the dictionary and might require a re-sorting of the related attribute vector, if the dictionary is sorted. Due to that, sequentially writing data to the database is not sped up, either.

5. Entropy

What is entropy?

Possible Answers:

(a) Entropy limits the amount of entries that can be inserted into a database. System specifications greatly affect this key indicator.
(b) Entropy represents the amount of information in a given dataset. It can be calculated as the number of distinct values in a column (column cardinality) divided by the number of rows of the table (table cardinality).
(c) Entropy determines tuple lifetime. It is calculated as the number of duplicates divided by the number of distinct values in a column (column cardinality).
(d) Entropy limits the attribute sizes. It is calculated as the size of a value in bits divided by number of distinct values in a column the number of distinct values in a column (column cardinality).

Correct Answer: (b)

Explanation: As in information theory, in this context, entropy determines the amount of information content gained from the evaluation of a certain message, in this case the dataset.

Compression

1. Sorting Compressed Tables

Which of the following statements is correct?

Possible Answers:

(a) If you sort a table by the amount of data for a row, you achieve faster read access

(b) Sorting has no effect on possible compression algorithms

(c) You can sort a table by multiple columns at the same time

(d) You can sort a table only by one column

Correct Answer: (d)

Explanation: Some compression techniques achieve a better compression rate when they are applied to a sorted table, like Indirect Encoding. Furthermore, a table cannot be sorted by multiple columns at the same time, so the right answer is that a table can only be sorted by one column without auxiliary data structures or positional information per attribute value. A potential enhancement is that one can sort a table cascading (i.e. first sort by country, then sort the resulting groups by city, ...), which improves accesses to the column used as secondary sorting attribute to some extent.

2. Compression and OLAP / OLTP

What do you have to keep in mind if you want to bring OLAP and OLTP together?

Possible Answers:

(a) You should not use any compression techniques because they increase CPU load

(b) You should not use compression techniques with direct access, because they cause major security concerns

(c) Legal issues may prohibit to bring certain OLTP and OLAP datasets together, so all entries have to be reviewed

(d) You should use compression techniques that give you direct positional access, since indirect access is too slow

Correct Answer: (d)

Explanation: Direct positional access is always favorable. It does not cause any difference to data security. Also, legal issues will not interfere with OLTP and OLAP datasets, since all OLAP data is generated out of OLTP data. The increased CPU load which occurs when using compression is tolerated because the compression leads to smaller data sizes which usually results in better cache usage and faster response times.

3. Compression Techniques for Dictionaries

Which of the following compression techniques can be used to decrease the size of a sorted dictionary?

Possible Answers:

(a) Cluster Encoding
(b) Prefix Encoding
(c) Run-Length Encoding
(d) Delta Encoding

Correct Answer: (d)
Explanation: Delta Encoding for Dictionaries is explained in detail in Sect. 7.5. Cluster Encoding, Run-Length Encoding, and Prefix Encoding can not be used on dictionaries because each dictionary entry is unique.

4. **Compression Example Prefix Encoding**
Suppose there is a table where all 80 million inhabitants of Germany are assigned to their cities. Germany consists of about 12,200 cities, so the valueID is represented in the dictionary via 14 bit. The outcome of this is that the attribute vector for the cities has a size of 140 MB. We compress this attribute vector with Prefix Encoding and use Berlin, which has nearly 4 million inhabitants, as the prefix value. What is the size of the compressed attribute vector?
Assume that the needed space to store the amount of prefix values and the prefix value itself is neglectable, because the prefix value only consumes 22 bit to represent the number of citizens in Berlin and additional 14 bit to store the key for Berlin once. Further assume the following conversions: $1\,MB = 1,000\,kB$, $1\,kB = 1,000\,B$
Possible Answers:

(a) $0.1\,MB$
(b) $133\,MB$
(c) $63\,MB$
(d) $90\,MB$

Correct Answer: (b)
Explanation: Because we use Prefix Encoding for this attribute vector, we do not save the valueID for Berlin 4 million times to represent its inhabitants in the city column. Instead, we resort the table so that all people who live in Berlin are on the top. In the attribute vector we save the valueID for Berlin and the number of occurrences for that valueID. Then the valueIDs for the remaining 76 million people in Germany follow. So the new size of the attribute vector is made up of the size of the valueID for Berlin (14 bit), the size needed to save the number of occurrences for Berlin (22 bit) and the size of the remaining entries. The missing numbers can be calculated the following way: From 80 million people in Germany remain 76 million to store. Each entry needs 14 bit for the valueID of its city. So the size of the remaining entries is 76 million $\cdot$ 14 bit $= 1,064,000,000$ bit. Thus, the size of the attribute vector is 14 bit $+ 22$ bit $+ 1,064,000,000$ bit $= 1,064,000,036$ bit, which is about 133 MB (8 bit $= 1$ byte).

5. Compression Example Run-Length Encoding Germany

Suppose there is a table where all 80 million inhabitants of Germany are assigned to their cities. The table is sorted by city. Germany consists of about 12,200 cities (represented by 14 bit). Using Run-Length Encoding with a start position vector, what is the size of the compressed city vector? Always use the minimal number of bits required for any of the values you have to choose and include all needed auxiliary structures. Further assume the following conversions: $1 \text{ MB} = 1{,}000 \text{ kB}$, $1 \text{ kB} = 1{,}000 \text{ B}$

Possible Answers:

(a) 1.2 MB
(b) 127 MB
(c) 5.2 kB
(d) 62.5 kB

Correct Answer: (d)

Explanation: We have to compute the size of (a) the value array and (b) the size of the start position array. The size of (a) is the distinct number of cities (12,200) times the size of each field of the value array ($\lceil log_2(12,200) \rceil$). The size of (b) is the number of entries in the dictionary (12,200) times the number of bit required to encode the highest possible number of inhabitants ($\lceil log_2(80,000,000) \rceil$). The result is thus 14 bit times 12,200 (170,800) plus 27 bit times 12,200 (329,400), summing up to 500,200 bit (or 62.5 kB) in total.

6. Compression Example Cluster Encoding

Assume the world population table with 8 billion entries. This table is sorted by countries. There are about 200 countries in the world. What is the size of the attribute vector for countries if you use Cluster Encoding with 1,024 elements per block assuming one block per country can not be compressed? Use the minimum required count of bits for the values and include all needed auxiliary structures. Further assume the following conversions: $1 \text{ MB} = 1{,}000 \text{ kB}$, $1 \text{ kB} = 1{,}000 \text{ B}$

Possible Answers:

(a) ≈ 9 MB
(b) ≈ 4 MB
(c) ≈ 0.5 MB
(d) ≈ 110 MB

Correct Answer: (a)

Explanation: To represent the 200 cities, 8 bit are needed for the valueID, because $\lceil log_2(200) \rceil$ is 8. With a cluster size of 1,024 elements the number of blocks is 7,812,500 (8 billion entries / 1,024 elements per block). Each country has one incompressible block, so there are 200 of them. The size of one incompressible block is the number of elements per block (1,024) times the size of one valueID (8 bit). The result is 8,192 bit and consequently the required size for the 200 blocks is $200 \cdot 8,192 \text{ bit} = 1,638,400$ bit. For the remaining 7,812,300 compressible blocks, it is only necessary to store one valueID for each block. Hence the resulting size of the compressible blocks is 62,498,400 bit ($7,812,300 \cdot 8$ bit). Finally there is the bit vector which indicates compressible and incompressible blocks. It requires 1 bit per block, so it has a size of 7,812,500 bit. The size of the whole compressed attribute vector is the sum of the size of the compressed and uncompressed blocks and the bit vector, which is $1,638,400 \text{ bit} + 62,498,400 \text{ bit} + 7,812,500 \text{ bit} = 71,949,300 \text{ bit}$. That is about 9 MB, which is the correct answer.

7. Best Compression Technique for Example Table
Find the best compression technique for the name column in the following table. The table lists the names of all inhabitants of Germany and their cities, i.e. there are two columns: first_name and city. Germany has about 80 million inhabitants and 12,200 cities. The table is sorted by the city column. Assume that any subset of 1,024 citizens contains at most 200 different first names.
Possible Answers:

(a) Run-Length Encoding
(b) Indirect Encoding
(c) Prefix Encoding
(d) Cluster Encoding

Correct Answer: (b)
Explanation: In order to use Prefix Encoding or Run-Length Encoding to compress a column, a table should be sorted by this specific column. In this example, we want to compress the "first_name" column, but the table is sorted by the column city. Therefore we can not use these two compression techniques. Cluster Encoding is possible in general and we could achieve high compression rates, but unfortunately Cluster Encoding does not support direct access. So choosing Cluster Encoding would prohibit direct access and consequently we would loose performance. As a conclusion, Indirect Encoding is the best compression technique for this column, because it works with a good compression rate while keeping the possibility of direct access.

Data Layout in Main Memory

1. Consecutive Access vs. Stride Access
When DRAM can be accessed randomly with the same costs, why are consecutive accesses usually faster than stride accesses?
Possible Answers:

(a) With consecutive memory locations, the probability that the next requested location has already been loaded in the cache is higher than with randomized/strided access. Furthermore is the memory page for consecutive accesses probably already in the TLB.

(b) The bigger the size of the stride, the higher the probability, that two values are both in one cache line.
(c) Loading consecutive locations is not faster, since the CPU performs better on prefetching random locations, than prefetching consecutive locations.
(d) With modern CPU technologies like TLBs, caches and prefetching, all three access methods expose the same performance.

Correct Answer: (a)

Explanation: Having always the same distance between accessed addresses enables the prefetcher to predict the correct locations to load. For randomly accessed addresses this is obviously not possible. Furthermore, strides of zero, which is the case for consecutive attribute accesses using columnar layouts, are highly cache efficient, since solely the needed locations are loaded. To summarize, random memory accesses might have the same costs, but since the CPU loads more data than requested into the caches and the prefetcher even fetches unrequested data, datasets of which the dictionaries exceed the cache size can be processed faster through consecutive accesses.

Partitioning

1. Partitioning Types
Which partitioning types do really exist and are mentioned in the course?
Possible Answers:

(a) Selective Partitioning
(b) Syntactic Partitioning
(c) Range Partitioning
(d) Block Partitioning

Correct Answer: (c)

Explanation: Range Partitioning is the only answer that really exists. It is a subtype of horizontal partitioning and separates tables into partitions by a predefined partitioning key, which determines how individual data rows are distributed to different partitions.

2. Partitioning Type for Given Query
Which partitioning type fits best for the column 'birthday' in the world population table, when we assume that the main workload is caused by queries like 'SELECT first_name, last_name FROM population WHERE birthday > 01.01.1990 AND birthday < 31.12.2010 AND country = 'England'? Assume a non-parallel setting, so we can not scan partitions in parallel. The only parameter that is changed in the query is the country.

Possible Answers:

(a) Round Robin Partitioning
(b) All partitioning types will show the same performance
(c) Range Partitioning
(d) Hash Partitioning

Correct Answer: (c)

Explanation: Range Partitioning separates tables into partitions by a predefined key. In the example that would lead to a distribution where all required tuples are in the same partition (or in the minimal number of partitions to cover the queried range) and our query only needs to access this (or these). Round Robin Partitioning would not be a good partitioning type in this example, because it assigns tuples turn by turn to each partition, so the data is separated across many different partitions which have to be accessed. Hash Partitioning uses a hash function to specify the partition assignment for each row, so the data is probably separated across many different partitions, too.

3. **Partitioning Strategy for Load Balancing**
   Which partitioning type is suited best to achieve fair load-balancing if the values of the column are non-uniformly distributed?
   Possible Answers:

   (a) Partitioning based on the number of attributes used modulo the number of systems
   (b) Range Partitioning
   (c) Round Robin Partitioning
   (d) All partitioning types will show the same performance

   Correct Answer: (c)

   Explanation: Round Robin Partitioning distributes tuples turn by turn to each partition, so all partitions have nearly the same number of tuples. In contrast, Range Partitioning assigns entries to the table by a predefined partitioning key. Because the values used as partitioning keys are normally distributed non-uniformly, it is difficult or maybe even impossible to find a key that segments the table into parts of the same size. Consequently, Round Robin Partitioning is the best strategy for fair load-balancing if the values of the column are non-uniformly distributed.

Delete

1. **Delete Implementations**
   Which two possible delete implementations are mentioned in the course?

Possible Answers:

(a) White box and black box delete
(b) Physical and logical delete
(c) Shifted and liquid delete
(d) Column and row deletes

Correct Answer: (b)
Explanation: A physical delete erases the tuple content from memory, so that it is no longer there. A logical delete only marks the tuple as invalid, but it may still be queried for history traces.

2. Arrays to Scan for Specific Query with Dictionary Encoding
When applying a delete with two predicates, e.g. firstname='John' AND lastname='Smith', how many logical blocks in the IMDB are being looked at during determination which tuples to delete (all columns are dictionary encoded)?
Possible Answers:

(a) 1
(b) 2
(c) 4
(d) 8

Correct Answer: (c)
Explanation: First the two dictionaries for firstname and lastname to get the corresponding valueIDs and then the two attribute vectors to get the positions (recordIDs).

3. Fast Delete Execution
Assume a physical delete implementation and the following two SQL statements on our world population table:

(A) DELETE FROM world_population WHERE country='China';
(B) DELETE FROM world_population WHERE country='Ireland';

Which query will execute faster? Please only consider the concepts learned so far.
Possible Answers:

(a) Equal execution time
(b) A
(c) Depends on the ordering of the dictionary
(d) B

Correct Answer: (d)
Explanation: Based on the actual locations of used logical blocks in a physical delete implementation, the largest part of the time will be moving memory blocks. Therefore the number of deletes is essential for the runtime. Since China has a much larger population than Ireland, query B will be faster.

Insert

1. Access Order of Structures During Insert
When doing an insert, what entity is accessed first?
Possible Answers:

(a) The attribute vector
(b) The dictionary
(c) No access of either entity is needed for an insert
(d) Both are accessed in parallel in order to speed up the process

Correct Answer: (b)
Explanation: First the dictionary is scanned in order to figure out, whether the value to be inserted is already part of the dictionary or has to be added.

2. New Value in Dictionary
Given the following entities:
Old dictionary: ape, dog, elephant, giraffe
Old attribute vector: 0, 3, 0, 1, 2, 3, 3
Value to be inserted: lamb
What value is the lamb mapped to in the new attribute vector?
Possible Answers:

(a) 1
(b) 2
(c) 3
(d) 4

Correct Answer: (d)
Explanation: "lamb" starts with the letter "l" and therefore does belong after the entry "giraffe". "giraffe" was the last entry with the logical number 3 (fourth entry) in the old dictionary, so "lamb" gets the number 4 in the new dictionary.

3. Insert Performance Variation over Time
Why might real world productive column stores experience faster insert performance over time?
Possible Answers:

(a) Because the dictionary reaches a state of saturation and, thus, rewrites of the attribute vector become less likely.
(b) Because the hardware will run faster after some run-in time.
(c) Because the column is already loaded into main-memory and does not have to be loaded from disk.
(d) An increase in insert performance should not be expected.

Correct Answer: (a)

Explanation: Consider for instance a database for the world population. Most first names probably did appear after writing a third of the world population. Future inserts can be done a little faster since less steps are required if the values are already present in the corresponding dictionary.

4. **Resorting Dictionaries of Columns**
   Consider a dictionary encoded column store (without a differential buffer) and the following SQL statements on an initially empty table:
   INSERT INTO students VALUES('Daniel', 'Bones', 'USA');
   INSERT INTO students VALUES('Brad', 'Davis', 'USA');
   INSERT INTO students VALUES('Hans', 'Pohlmann', 'GER');
   INSERT INTO students VALUES('Martin', 'Moore', 'USA');
   How often do attribute vectors have to be completely rewritten?
   Possible Answers:

   (a) 2
   (b) 3
   (c) 4
   (d) 5

Correct Answer: (b)
Explanation: Each column needs to looked at separately. An attribute vector always gets rewritten, if the dictionary was resorted.

- First name: 'Daniel' gets inserted, its the first dictionary entry. When 'Brad' gets added to the dictionary, it needs to be resorted and therefore the attribute vector needs to be rewritten. 'Hans' and 'Martin' are simply appended to the end of the dictionary each time. Thats a total of one rewrite for the first name.

For the other attributes, the process is equal, the actions are described in short:

- Last name: Bones, Davis, Pohlmann, Moore → rewrite
- Country: USA, USA → already present, GER → rewrite, USA → already present

In total, three rewrites are necessary.
Insert Performance
Which of the following use cases will have the worst insert performance when all values will be dictionary encoded?
Possible Answers:

(a) A city resident database, that store all the names of all the people from that city
(b) A database for vehicle maintenance data which stores failures, error codes and conducted repairs
(c) A password database that stores the password hashes
(d) An inventory database of a company storing the furniture for each room

Correct Answer: (c)

Explanation: Inserts take especially long when new unique dictionary entries are inserted. This will be most likely the case for time stamps and password hashes.

Update

1. Status Update Realization
How do we want to realize status updates for binary status variables?
Possible Answers:

(a) Single status field: "false" means state 1, "true" means state 2
(b) Two status fields: "true/false" means state 1, "false/true" means state 2
(c) Single status field: "null" means state 1, a timestamp signifies transition to state 2
(d) Single status field: timestamp 1 means state 1, timestamp 2 means state 2

Correct Answer: (c)

Explanation: By using "null" for state 1 and a timestamp for state 2, the maximum density of needed information is achieved. Given a binary status, the creation time of the initial status is available in the creation timestamp of the complete tuple, it does not have to be stored again. If the binary information is flipped, the "update" timestamp conserves all necessary information in the described manner. Just saving "true" or "false" would discard this information.

2. Value Updates
What is a "value update"?
Possible Answers:

(a) Changing the value of an attribute
(b) Changing the value of a materialized aggregate
(c) The addition of a new column
(d) Changing the value of a status variable

Correct Answer: (a)

Explanation: In typical enterprise applications, three different types of updates can be found. Aggregate updates change a value of a materialized aggregate, status updates change the value of a status variable and finally value updates change the value of an attribute. Adding a new, empty column is not regarded as an update of a tuple at all, because it manipulates the whole relation via the database schema.

3. Attribute Vector Rewriting After Updates
Consider the world population table (first name, last name) that includes all people in the world: Angela Mueller marries Friedrich Schulze and becomes Angela Schulze. Should the complete attribute vector for the last name column be rewritten?

Possible Answers:

(a) No, because 'Schulze' is already in the dictionary and only the valueID in the respective row will be replaced
(b) Yes, because 'Schulze' is moved to a different position in the dictionary
(c) It depends on the position: All values after the updated row need to be rewritten
(d) Yes, because after each update, all attribute vectors affected by the update are rewritten

Correct Answer: (a)

Explanation: Because the entry 'Schulze' is already in the dictionary, it implicitly has the correct position concerning the sort order and does not need to be moved. Furthermore, each attribute vector entry has a fixed size, so that every dictionary entry can be referenced without changing the position of the adjacent entries in the memory area. Based on these two facts, the answer is that the attribute vector does not need to be rewritten.

Tuple Reconstruction

1. Tuple Reconstruction on the Row Layout: Performance

Given a table with the following characteristics:

- Physical storage in rows
- The size of each field is 34 byte
- The number of attributes is 9
- A cache line has 64 byte
- The CPU processes 4 MB/ms.

Calculate the time required for reconstructing a full row. Please assume the following conversions: $1\,MB = 1{,}000\,kB$, $1\,kB = 1{,}000\,B$
Possible Answers:

(a) $\approx 0.05\,\mu s$
(b) $\approx 0.125\,\mu s$
(c) $\approx 0.08\,\mu s$
(d) $\approx 0.416\,\mu s$

Correct Answer: (c)

Explanation: For 9 attributes of each 34 byte rows, in total 306 byte have to be fetched. Given a cache line size of 64 byte, 5 cache lines have to be filled: $5 \cdot 64$ byte $= 320$ byte;

Total time needed: size of data to be read / processing speed $= 320$ byte / $4{,}000{,}000$ byte/ms/core $= 0.08\,\mu s$

2. Tuple Reconstruction on the Column Layout: Performance

Given a table with the following characteristics:

- Physical storage in columns
- The size of each field is 34 byte
- The number of attributes is 9
- A cache line has 64 byte
- The CPU processes 4 MB/ms

Calculate the time required for reconstructing a full row. Please assume the following conversions: $1\,MB = 1{,}000\,kB$, $1\,kB = 1{,}000\,B$

Possible Answers:

(a) $\approx 0.08\,\mu s$
(b) $\approx 0.725\,\mu s$
(c) $\approx 0.144\,\mu s$
(d) $\approx 0.225\,\mu s$

Correct Answer: (c)

Explanation: Size of data to be read: number of attributes $\cdot$ cache line size $= 9 \cdot 64 = 576$ byte

Total time needed: size of data to be read / processing speed $= 576$ byte / $4{,}000{,}000$ byte/ms/core $= 0.144\,\mu s$

3. Tuple Reconstruction in Hybrid Layout

A table containing product stock information has the following attributes:
Warehouse (4 byte); Product Id (4 byte); Product Name Short (20 byte); Product Name Long (40 byte); Self Production (1 byte); Production Plant (4 byte); Product Group (4 byte); Sector (4 byte); Stock Volume (8 byte); Unit of Measure (3 byte); Price (8 byte); Currency (3 byte); Total Stock Value (8 byte); Stock Currency (3 byte)

The size of a full tuple is 114 byte.
The size of a cache-line is 64 byte.
The table is stored in main memory using a hybrid layout. The following fields are stored together:

- Stock Volume and Unit of Measure;
- Price and Currency;
- Total Stock Value and Stock Currency;

All other fields are stored column-wise.
Calculate and select from the list below the time required for reconstructing a full tuple using a single CPU core with a scan speed of 4 MB/ms. Please assume the following conversions: $1\,MB = 1{,}000\,kB$, $1\,kB = 1{,}000\,B$

Possible Answers:

(a) $\approx 0.176\,\mu s$
(b) $\approx 0.04\,\mu s$

(c) $\approx 0.03\,\mu s$

(d) $\approx 0.213\,\mu s$

Correct Answer: (a)

Explanation: The correct answer is calculated as follows: first, the number of cache lines to be accessed is determined. Attributes that are stored together, can be read in one access if the total bit size to be retrieved does not exceed the size of a cache line.

Stock Volume and Unit of Measure: 8 byte + 3 byte < 64 byte → 1 cache line

Price and Currency: 8 byte + 3 byte < 64 byte → 1 cache line

Total Stock Value and Stock Currency: 8 byte + 3 byte < 64 byte → 1 cache line

All other 8 attributes are stored column wise, so one cache access per attribute is required, resulting in additional cache accesses.

The total amount of data to be read is therefore: 11 (cachelines) · 64 byte = 704 byte 704 byte / (4,000,000 byte/ms/core) · 1 core = 0.176 μs

4. **Comparison of Performance of the Tuple Reconstruction on Different Layouts**

A table containing product stock information has the following attributes:

Warehouse (4 byte); Product Id (4 byte); Product Name Short (20 byte); Product Name Long (40 byte); Self Production (1 byte); Production Plant (4 byte); Product group (4 byte); Sector (4 byte); Stock Volume (8 byte); Unit of Measure (3 byte); Price (8 byte); Currency (3 byte); Total Stock Value (8 byte); Stock Currency (3 byte)

The size of a full tuple is 114 byte.

The size of a cache-line is 64 byte.

The scan speed of a CPU core is 4 MB/ms.

Which of the following statements are true?

Possible Answers:

(a) If the table is physically stored in column layout, the reconstruction of a single full tuple consumes $\approx 0.096\,\mu s$ using a single CPU core.

(b) If the table is physically stored in row layout, the reconstruction of a single full tuple consumes $\approx 64\,ns$ using a single CPU core.

(c) If the table is physically stored in column layout, the reconstruction of a single full tuple consumes $\approx 224\,ns$ using a single CPU core.

(d) If the table is physically stored in row layout, the reconstruction of a single full tuple consumes $\approx 0.32\,\mu s$ using a single CPU core.

Correct Answer: (c)

Explanation: To reconstruct a full tuple from row layout, we first need to calculate the count of cache accesses. Considering a size of 114 byte, we will need 2 cache accesses (114 byte / 64 byte per cache line = 1.78 → 2) to read a whole tuple from main memory. With two cache accesses, each loading 64 byte, we read 128 byte from main memory. We assume, like in the questions before, that the reading speed of our system is 4 MB/ms/core. Now we divide 128 byte by 4 MB/ms/core and get a result of 0.000032 ms (0.032 μs). So the answers ≈ 0.32 μs and ≈ 64 ns with one core, if the table is stored in row layout, are both false.

In a columnar layout, we need to read every value individually from main memory, because they are not stored in a row (consecutive memory area) but in different attribute vectors. We have 14 attributes in this example, so we need 14 cache accesses, each reading 64 bytes. Thus the CPU has to read 14 · 64 byte = 896 byte from main memory. Like before, we assume that the reading speed is 4 MB/ms/core. By dividing the 896 byte by 4 MB/ms/core, we get the time one CPU core needs to reconstruct a full tuple. The result is 0.000224 ms (224 ns). So ≈ 224 ns with one core and columnar layout is the correct answer.

Scan Performance

1. **Loading Tuples from a Dictionary-Encoded Row-Oriented Layout**
Consider the example presented in Sect. 14.2. We will now do the calculation assuming we have dictionary-encoded attributes. For this question, let us assume that each of the 8 billion encoded tuples has a total size of 32 byte. What is the time that a single core processor with a scan speed of 4 MB/ms needs to scan the whole world_population table if all data is stored in a dictionary-encoded row layout?
Possible Answers:

(a) 128 s
(b) 32 s
(c) 48 s
(d) 64 s

Correct Answer: (d)
Explanation: The accessed data volume is 8 billion tuples · 32 byte each ≈ 256 GB. Therefore the expected response time is 256 GB / (4 MB/ms/core · 1 core) = 64 s

Select

1. **Optimizing SELECT**
How can the query optimizer improve the performance of SELECT statements?
Possible Answers:

(a) By reducing the number of indices
(b) By using the FAST SELECT keyword

(c) Optimizers try to keep intermediate result sets large for maximum flexibility during query processing
(d) By ordering multiple sequential select statements from strong (low) selectivity to weak (high) selectivity

Correct Answer: (d)
Explanation: During the execution of the SELECT statement, we have to search through the whole data of the database for entries with the demanded selection attributes. By ordering the selection criteria from strong/low (many rows are filtered out) to weak/high selectivity (few rows are filtered out), we reduce the amount of data we have to walk through in subsequent steps, which results in a shorter execution time for the overall SELECT statement.

2. **Selection Execution Order**
   Given is a query that selects the names of all German women born in the last 10 years from a world_population table, that stores information about all humans born in the last 100 years. In which order should the query optimizer execute the selections at best? Assume a sequential query execution plan and use the predicate selectivities for your calculations.
   Possible Answers:

   (a) gender first, country second, birthday last
   (b) country first, gender second, birthday last
   (c) country first, birthday second, gender last
   (d) birthday first, gender second, country last

Correct Answer: (c)
Explanation: To optimize the speed of sequential selections the most restrictive ones have to be executed first. While the gender restriction would reduce the amount of data (8 billion) to the half (4 billion), the birthday restriction would return about 800 million tuples and the country restriction (Germany) about 80 million tuples. So the query optimizer should execute the country restriction first, followed by the birthday restriction, which filters additional 80 % of the 80 million Germans. The gender restriction then filters the last $\approx$ 50 % of entries. The query optimizer however does not have this fine grained data distribution information and has usually to rely on attribute selectivities. Based on that, the query optimizer would start with the birthday attribute, which has a selectivity of 1 / 36,500 (365 days per year · 100 years). The gender attribute has a selectivity of 1/2 and the country attribute has a selectivity of 1 / 200 (if we assume 200 countries). Because our query selects all women born in the last 10 years, the actual achieved tuple reduction by applying the birthday predicate is only about 1/10 instead of 1 / 36,500. This example illustrates the difference between predicate and attribute selectivity. It shows furthermore, that the best possible execution order is hard to be determined just based on the attribute selectivities.

3. Selectivity Calculation

Given is the query to select the names from German men born after January 1, 1990 and before December 31, 2009 from the world population table (8 billion people). Calculate the selectivity.

Selectivity = number of tuples selected / number of tuples in the table

Assumptions:

- there are about 80 million Germans in the table
- males and females are equally distributed in each country (50/50)
- there is an equal distribution between all generations from 1910 until 2010

Possible Answers:

(a) 0.001
(b) 0.005
(c) 0.1
(d) 1

Correct Answer: (a)

Explanation: Calculation:

- 80 million Germans
- 80 million (Germans) $\cdot$ 50 % = 40 million German males
- 40 million (German males) $\cdot$ 20 % = 8 million German males between 1990 and 2010
- 8 million (German males between 1990 and 2010) / 8 billion = 0.001

The first selection is based on the assumption, that the distribution of males and females is equal. The second selection is based on the assumption that the population is equally distributed over the generations, so selecting a timespan of 20 years of 100 years is effectively 1/5th or 20 %. The selectivity is then calculated via: number of selected tuples / number of all tuples = 8 million / 8 billion = 0.001

Materialization Strategies

1. Performance of Materialization Strategies

Which materialization strategy—late or early materialization—provides the better performance?

Possible Answers:

(a) Depends on the characteristics of the executed query
(b) Late and early materialization always provide the same performance
(c) Late materialization
(d) Early materialization

Correct Answer: (a)

Explanation: The question of which materialization strategy to use is dependent on several facts. Amongst them are e.g. the selectivity of queries and the execution strategy (pipelined or parallel). In general, late materialization is superior, if the query has a low selectivity and the queried table uses compression.

2. Characteristics of Early Materialization

Which of the following statements is true?

Possible Answers:

(a) Whether late or early materialization is used is determined by the system clock
(b) Early materialization requires lookups into the dictionary, which can be expensive and are not required when using late materialization
(c) Depending on the persisted value types of a column, using positional information instead of actual values can be advantageous (e.g. in terms of cache usage or SIMD execution)
(d) The execution of an early materialized query plan can not be parallelized

Correct Answer: (c)

Explanation: Working with intermediate results provides advantages in terms of cache usage and parallel execution, since positional information usually has a smaller size than the actual values. Consequently, more items fit into a cache line, which additionally are of fixed length enabling parallel SIMD operations. The question of locking and parallelization is in general independent from the materialization strategy.

Parallel Data Processing

1. Amdahl's Law

Amdahl's Law states that ...

Possible Answers:

(a) the number of CPUs doubles every 18 months
(b) the amount of available memory doubles every year
(c) the speedup of parallelization is limited by the time needed for the sequential fractions of the program
(d) the level of parallelization can be no higher than the number of available CPUs

Correct Answer: (c)

Explanation: While the execution time of the parallelizable code segments can be shortened by multiple cores, the runtime of the sequential fraction can not be decreased because it has to be executed by one CPU core only. As this time is constant, the execution time of the complete program code is at least as long as the sequential code segment, regardless how many cores are used. This main principle was first described by Amdahl and named after him. The increase of chip density by a factor of 2 about every 18–24 month is called "Moore's Law", the other two possible answers are incorrect at all.

2. Shared Memory
What limits the use of shared memory?
Possible Answers:

(a) The operation frequency of the processor
(b) The usage of Streaming SIMD instructions (SSE).
(c) The caches of each CPU
(d) The number of worker threads, which share the same resources and the limited memory itself.

Correct Answer: (d)

Explanation: By default, main memory is assigned to exactly one process, all other process can not access the reserved memory area. If a memory segment is shared between processes and this segment is full, additional shared memory has to be requested. The size of the memory area is therefore a limiting factor. But more important is the fact that several workers (this can be threads, processes, etc.) share the same resource. To avoid inconsistencies, measures have to be taken, e.g. locking or MVCC (multiversion concurrency control). While this does usually not affect low numbers of workers, any locking will automatically get a problem as soon as the number of workers reaches a certain limit. At this level, too many workers cannot work since they have to wait for locked resources.

Indices

1. Index Characteristics
Introducing an index...
Possible Answers:

(a) increases memory consumption
(b) speeds up inserts
(c) slows down look-ups
(d) decreases memory consumption

Correct Answer: (a)

Explanation: An index is an additional data structure and therefore consumes memory. Its purpose is to increase performance of scan operations.

2. Inverted Index
What is an inverted index?
Possible Answers:

(a) A list of text entries that have to be decrypted. It is used for enhanced security.
(b) A structure that maps each distinct value to a position list, which contains all positions where the value can be found in the column
(c) A structure that contains the distinct values of the dictionary in reverse order
(d) A structure that contains the delta of each entry in comparison to the largest value

Correct Answer: (b)
Explanation: The inverted index consists of the index offset vector and the index position vector. The index offset vector stores a reference to the sequence of positions of each dictionary entry in the index position vector. The index position vector contains the positions (i.e. all occurrences) for each value in the attribute vector. Thus, the inverted index is a structure that maps each distinct value to a position list, which contains all positions where the value can be found in the column.

Join

1. Sort-Merge Join Complexity
What is the complexity of the Sort-Merge Join?
Possible Answers:

(a) $O(n \cdot m)$
(b) $O(n^2/m^2)$
(c) $O(n+m)$
(d) $O(n \cdot \log(n)+m \cdot \log(m))$

Correct Answer: (d)
Explanation: Let m and n be the cardinality of the input relations M and N with $m \leq n$. The runtime of the Sort-Merge Join is determined by the task to sort both input relations. As sorting algorithm, merge sort is used, that has runtime a complexity of $O(n \cdot \log(n))$ for input relation N. This complexity is based on the fact, that the merge join works recursive and divides the input into two parts, which are sorted and afterwards combined. The actual number of steps is determined by the number of recursion levels. The resulting equation for n elements can be assessed with the term $n \cdot \log(n)$ according to the master theorem. Therefore, the Sort-Merge Join has a complexity of $O(m \cdot \log(m)+n \cdot \log(n))$.

2. Join Algorithm for Small Data Sets

Given is an extremely small data set. Which join algorithm would you choose in order to get the best performance?

Possible Answers:

(a) Nested-Loop Join
(b) All join algorithms have the same performance
(c) Hash Join
(d) Sort-Merge Join

Correct Answer: (a)

Explanation: Even though the Nested-Loop Join has a much worse complexity than the other join algorithms, it manages to join the input relations without additional data structures. Therefore, no initialization is needed. In case of very small relations, this is a huge saving.

3. Equi Join

What is the Equi Join?

Possible Answers:

(a) It is a join algorithm to fetch information, that is probably not there. So if you select a tuple from one relation and this tuple has no matching tuple on the other relation, you would insert NULL values there.
(b) It is a join algorithm that ensures that the result consists of equal amounts from both joined relations
(c) If you select tuples from both relations, you use only one half of the join relations and the other half of the table is discarded
(d) If you select tuples from both relations, you will always select those tuples, that qualify according to a given equality predicate

Correct Answer: (d)

Explanation: There are two general categories of joins: inner joins and outer joins. Inner joins create a result table that combines tuples from both input tables only if both regarded tuples meet the specified join condition. Based on a join predicate each tuple from the first table is combined with each tuple of the second table, the resulting cross-set is filtered (similar to a SELECT statement) on the join condition. Outer joins, in contrast, have more relaxed conditions on which tuples to include. If a tuple has no matching tuple in the other relation, the outer join inserts NULL values for the missing attributes in the result and includes the resulting combination. Further specializations of the two join types are for example the Semi-Join, which returns only the attributes of the left join partner if the join predicate is matched. Another specialization is the Equi-Join. It allows the retrieval of tuples that satisfy a given equality predicate for both sides of the tables to be joined. The combined result comprises the equal attribute twice, once from the left relation and once from the right relation. The so called Natural-Join is similar to the Equi-Join, except that it strips away the redundant column that is kept in the Equi-Join.

4. One-to-One-Relationship
What is a one-to-one relationship?
Possible Answers:

(a) Each query which has exactly one join between exactly two tables is called a one-to-one relationship, because one table is joined to exactly one other table.
(b) A one-to-one relationship between two objects means that for exactly one object on the left side of the join exists exactly zero or one object on the right side and vice versa
(c) A one-to-one relationship between two objects means that for each object on the left side, there are one or more objects on the right side of the joined table and each object of the right side has exactly one join partner on the left
(d) A one-to-one relationship between two objects means that each object on the left side is joined to one or more objects on the right side of the table and vice versa each object on the right side has one or more join partners on the left side of the table

Correct Answer: (b)
Explanation: Three different types of relations exist between two tables. These are one-to-one, one-to-many and many-to-many relationships. In a many-to-many relationship, each tuple from the first relation may be related to multiple tuples from the second one and vice versa. In a one-to-many relationship, each tuple of the first table might be related to multiple tuples of the second table, but each tuple of the second relation refers to exactly one tuple of the first relation. Consequently a one-to-one relationship connects each tuple of the first relation with exactly zero or one tuple from the second relation and vice versa.

5. Join Algorithms
Which of these Join algorithms exists?
Possible Answers:

(a) Reverse-Traversal Join
(b) Bubble Join
(c) Bootstrap Join
(d) Nested-Loop Join

Correct Answer: (d)
Explanation: The Nested-Loop-Join cycles through all combinations of tuples from the involved relations by iterating through an interlaced loop per input relation.

Aggregate Functions

1. Aggregate Function Definition
What are aggregate functions?

Possible Answers:

(a) A set of indexes that speed up processing a specific report
(b) A specific set of functions that summarize multiple rows from an input data set
(c) A set of functions that transform data types from one to another data type
(d) A set of tuples that are grouped together according to specific requirements

Correct Answer: (b)
Explanation: Sometimes it is necessary to get a summary of a data set, like the average, the minimum, or the number of entries. Databases provide special functions for these tasks, called aggregate functions, that take multiple rows as an input and create an aggregated output. Instead of processing single values, these functions work on data sets, which are created by grouping the input data on specified grouping attributes.

2. Aggregate Functions

Which of the following is an aggregate function?
Possible Answers:

(a) GROUP BY
(b) MINIMUM
(c) HAVING
(d) SORT

Correct Answer: (b)
Explanation: MINIMUM (often expressed as MIN) is the only aggregate function listed here. HAVING is used in SQL Queries to add additional requirements for the resulting aggregate values in order to be accepted. GROUP BY is used to specify the columns on which the aggregations should take place (all tuples with equal values in this columns are grouped together, if multiple columns are specified, one result per unique attribute combination is computed). SORT is not a valid SQL expression at all.

Parallel Select

1. Query Execution Plans in Parallelizing SELECTS

When a SELECT statement should be executed in parallel ...
Possible Answers:

(a) all other SELECT statements have to be paused
(b) its query execution plan is not changed at all
(c) its query execution plan becomes much simpler compared to sequential execution
(d) its query execution plan has to be adapted accordingly

Correct Answer: (d)

Explanation: If columns are split into chunks, attribute vector scans can be executed in parallel with one thread per chunk. All results for chunks of the same column have to be combined by a UNION operation afterwards. If more than one column is scanned, a positional AND operation has to be performed additionally. In order to execute the AND operation in parallel too, the affected columns have to be partitioned equally. So parallelizing the AND operation changes the execution plan, therefore the answer that the execution plan has to be adapted, is correct.

Workload Management and Scheduling

1. Resource Conflicts
Which three hardware resources are usually taken into account by the scheduler in a distributed in-memory database setup?
Possible Answers:

(a) Main memory, disk and tape drive sizes
(b) CPU processing power, main memory size, network bandwidth
(c) Network bandwidth, power supply unit, main memory
(d) CPU processing power, graphics card performance, monitor resolution

Correct Answer: (b)
Explanation: When scheduling queries in an in-memory database, storage outside of main memory is of no importance. Furthermore, graphics hardware and peripherals are no performance indicator for database systems.

2. Workload Management Scheduling Strategy
Why does a complex workload scheduling strategy might have disadvantages in comparison to a simple resource allocation based on heuristics or a uniform distribution, e.g. Round Robin?
Possible Answers:

(a) A scheduling strategy is based on general workloads and thus might not reach the best performance for specific workloads compared to heuristics or a uniform distribution, while its application is cheap.
(b) Heuristics are always better than complex scheduling strategies.
(c) The execution of a scheduling strategy itself consumes more resources than a simplistic scheduling approach. A strategy is usually optimized for a certain workload—if this workload changes abruptly, the scheduling strategy might perform worse than a uniform distribution.
(d) Round-Robin is usually the best scheduling strategy.

Correct Answer: (c)

Explanation: Scheduling strategies reach a point where every further optimization is based on a specific workload. If one can predict future workloads based on the past ones, the scheduler can distribute queries for maximum performance regarding this scenario. However, if the workload changes unpredictably, there is no gain from specialized strategies. Under these circumstances, specialized optimizations are rather an unnecessary overhead, as they may require additional scheduling time, data structures, or increased resources.

3. Analytical Queries in Workload Management

Analytical queries typically are ...

Possible Answers:

(a) short running with soft time constraints
(b) short running with strict time constraints
(c) long running with soft time constraints
(d) long running with strict time constraints

Correct Answer: (c)

Explanation: Analytical workloads consist of complex and computationally heavy queries. Hence analytical queries have a long execution time. Whereas the response time must be guaranteed for transactional queries, this is not the case for analytical ones. While they should be as short as possible, business processes will not abort if an analytical query takes 3 instead of 2 s. Therefore, analytical queries have soft time constraints in comparison to transactional ones.

4. Query Response Times

Query response times ...

Possible Answers:

(a) have to be as short as possible, so the user stays focused at the task at hand
(b) can be increased so the user can do as many tasks as possible in parallel because context switches are cheap
(c) have no impact on a users work behavior
(d) should never be decreased as users are unfamiliar with such system behavior and can become frustrated

Correct Answer: (a)

Explanation: Query response times have a huge impact on the user. If an operation takes too long, the mind tends to wander to other topics than the original task. The longer it takes, the further this process goes on. Refocusing on the task is in fact very exhausting and thus, waiting periods should be avoided to guaranty a convenient experience and reduce human errors.

Parallel Join

1. Parallelizing Hash-Join Phases

What is the disadvantage when the probing phase of a join algorithm is parallelized and the hashing phase is performed sequentially?

Possible Answers:

(a) The algorithm still has a large sequential part that limits its potential to scale
(b) Sequentially performing the hashing phase introduces inconsistencies in the produced hash values
(c) The table has to be split into smaller parts, so that every core, which performs the probing, can finish
(d) The sequential hashing phase will run slower due to the large resource utilization of the parallel probing phase

Correct Answer: (a)

Explanation: With Amdahls' Law in mind, the Hash-Join can only be as fast as the sum of all sequential parts of the algorithm. Parallelizing the probing phase shortens the time needed, but has no impact on the hashing phase, which still has to be done sequentially. Thus, there is always a huge part of the algorithm which cannot be parallelized.

Parallel Aggregation

1. Aggregation: GROUP BY

Assume a query that returns the number of citizens of a country, e.g.:
SELECT country, COUNT(∗)
FROM world_population
GROUP BY country;
The world_population table contains the names and countries of all citizens of the world.
The GROUP BY clause is used to express . . .

Possible Answers:

(a) that the aggregate function shall be computed for every distinct value of country
(b) the graphical format of the results for display
(c) an additional filter criteria based on an aggregate function
(d) the sort order of countries in the result set

Correct Answer: (a)

Explanation: In general, the GROUP BY clause is used to express that all used aggregate function shall be computed for every distinct value (or value combinations) of the specified attributes. In this case, only one attribute is specified, so only sets having distinct values for country are aggregated. The sort order is specified in the ORDER BY clause, additional filter criteria can be added on the aggregated values with the HAVING clause.

2. Number of Threads
How many threads will be used during the second (aggregation) phase of the described parallel aggregation algorithm when the table is split into 20 chunks and the GROUP BY attribute has 6 distinct values?
Possible Answers:

(a) at least 10 threads
(b) at most 6 threads
(c) exactly 20 threads
(d) at most 20 threads

Correct Answer: (b)
Explanation: In the aggregation phase, the so called merger threads merge the buffered hash tables. Each thread is responsible for a certain range of the GROUP BY attribute. So if the GROUP BY attribute has 6 distinct values, the maximum number of threads is 6. If there are more than 6 threads, the surplus threads will get no pending values and are therefore not used.

Differential Buffer

1. The Differential Buffer
What is the differential buffer?
Possible Answers:

(a) A buffer where different results for one and the same query are stored for later usage
(b) A dedicated storage area in the database where inserts, updates and deletes are buffered
(c) A buffer where exceptions and error messages are stored
(d) A buffer where queries are stored until there is an idle CPU available for processing

Correct Answer: (b)
Explanation: The main store is optimized for read operations. An insert of a tuple is likely to force restructuring of the whole table. To avoid this, we introduced the differential buffer, an additional storage area where all the data modifications like inserts, updates and delete operations are performed and buffered until they are integrated into the main store. As a result, we have a read optimized main store and a write optimized differential buffer. In combination, update operations as well as read operations are supported by optimal storage structures, which results in an increased overall performance.

2. Performance of the Differential Buffer

Why might the performance of read queries decrease, if a differential buffer is used?

Possible Answers:

(a) Because only one query at a time can be answered by using the differential buffer

(b) Because the CPU cannot perform the query before the differential buffer is full

(c) Because read queries have to query both the main store and the write-optimized differential buffer

(d) Because inserts collected in the differential buffer have to be merged into the main store every time a read query comes in

Correct Answer: (c)

Explanation: New tuples are inserted into the differential buffer first, before being merged into the main store eventually. To speed up inserts as much as possible, the differential buffer is optimized for writing rather than reading tuples. Read queries against all data have to go against the differential buffer as well, what might cause a slowdown. To prevent noticeable effects for the user, the amount of values in the differential buffer is kept small in comparison to the main store. By exploiting the fact that the main store and the differential buffer can be scanned in parallel, noticeable speed losses are avoided.

3. Querying the Differential Buffer

If we use a differential buffer, we have the problem that several tuples belonging to one real world entry might be present in the main store as well as in the differential buffer. How did we solve this problem?

Possible Answers:

(a) This statement is completely wrong because multiple tuples for one real world entry never exist

(b) We introduced a validity bit

(c) We use a specialized garbage collector that just keeps the most recent entry

(d) All attributes of every doubled occurrence are set to NULL in the compressed main store.

Correct Answer: (b)

Explanation: Following the insert-only approach, we do not delete or change existing attributes or whole tuples. If we want to change or add attributes of an existing tuple in spite of the insert-only approach, we add an updated tuple with the changed as well as the unchanged values to the differential buffer. In order to solve the problem of multiple tuples belonging to one real world entry, we introduced a validity vector that indicates whether a tuple is the most current one and therefore is valid or not.

Insert Only

1. Statements Concerning Insert-Only
Considering an insert-only approach, which of the following statements is true?
Possible Answers:

(a) Using an Insert-only approach, invalidated tuples can no longer be used for time travel queries
(b) Old data items are deleted as they are not necessary any longer
(c) Data is not deleted, but invalidated instead
(d) Historical data has to be stored in a separate database to reduce the overall database size

Correct Answer: (c)
Explanation: With an insert-only approach, no data is ever deleted or sourced out to another, separated database. Furthermore, when using a differential buffer, the insert performance is not dependent on the data already stored in the database. Without the differential buffer, a huge amount of data already in the table might indeed speed up inserts into sorted columns, because a quite saturated and therefore stable dictionary would reduce the resorting overhead.

2. Benefits of Historical Data
Consider keeping deprecated or invalidated (i.e. historical) tuples in the database. Which of the following statements is wrong?
Possible Answers:

(a) It is legally required in many countries to store historical data
(b) Analyses of historical data can be used to improve the scan performance.
(c) Historical data can provide snapshots of the database at certain points in time
(d) Historic data can be used to analyze the development of the company

Correct Answer: (b)
Explanation: With historical data, time-travel queries are possible. They allow users to see the data exactly like it was at any point in the past. This simple access to historical data helps a company's management to efficiently analyze the history and the development of the enterprise. Additionally, i.e. in Germany it is legally required to store particular commercial documents for tax audits. Historical data however will not improve the scan performance, which is the correct answer consequently.

3. Accesses for Point Representation
Considering point representation and a table with one tuple, that was invalidated five times, how many tuples have to be checked to find the most recent tuple?
Possible Answers:

(a) Two, the most recent one and the one before that
(b) Five

(c) Six

(d) Only one, that is, the first which was inserted

Correct Answer: (c)

Explanation: All tuples belonging to one real world entry have to be checked in order to determine the most recent tuple when using point representation. It is not sufficient to start from the end of the table and take the first entry that belongs to the desired real world entry, because in most cases the table is sorted by an attribute and not by the insertion order. In this case, six tuples have to be checked, all five invalidated ones and the current one.

4. Statement Concerning Insert-Only

Which of the following statements concerning insert-only is true?

Possible Answers:

(a) Interval representation allows more efficient write operations than point representation

(b) In interval representation, four operations have to be executed to invalidate a tuple

(c) Point representation allows faster read operations than interval representation due to its lower impact on tuple size

(d) Point representation allows more efficient write operations than interval representation

Correct Answer: (d)

Explanation: Point representation will be less efficient for read operations, that only require the most recent tuple. Using point representation, all tuples of that entry have to be checked, to determine the most recent one. As a positive aspect, point representation allows more efficient write operations in comparison to interval representation, because on any update, only the tuple with the new values and the current 'value from' date has to be entered, the other tuples do not need to be changed. The insertion of the tuple with the new entries might however require the lookup of the former most recent tuple, to retrieve all unchanged values.

Merge

1. What is the Merge?

The merge process ...

Possible Answers:

(a) combines the main store and the differential buffer to increase the parallelism

(b) merges the columns of a table into a row-oriented format

(c) optimizes the write-performance

(d) incorporates the data of the write-optimized differential buffer into the read-optimized main store

Correct Answer: (d)

Explanation: If we are using a differential buffer as an additional data structure to improve the write performance of our database, we have to integrate this data into the compressed main partition periodically in order to uphold the benefits concerning the read performance. This process is called "merge".

2. When to Merge?

When is the merge process triggered?

Possible Answers:

(a) Before each SELECT operation

(b) When the space on disk runs low and the main store needs to be further compressed

(c) After each INSERT operation

(d) When the number of tuples within the differential buffer exceeds a specified threshold

Correct Answer: (d)

Explanation: Holding too many tuples in the differential buffer, slows down read performance against all data. Therefore it is necessary to define a certain threshold at which the merge process is triggered. If it was too often (for example after every INSERT or even before every SELECT), the overhead of the merge would be larger than the possible penalty for querying both main store and differential buffer.

Aggregate Cache

1. Aggregate Entries

What does the aggregate cache store in the aggregate entries?

Possible Answers:

(a) Aggregate query results computed on the main storage

(b) Aggregate query results computed on the main storage and the differential buffer

(c) Any type of query result

(d) Aggregate query results computed on the differential buffer

Correct Answer: (a)

Explanation: The aggregate cache only stores the results from queries on the main storage. On every cache hit, it triggers an on-the-fly aggregation on the differential buffer and combines the returning result with the aggregate entry in order to provide an up-to-date result.

2. Metric Entries
For what purpose is the information in the metric entries mainly used?
Possible Answers:

(a) Aggregations on the differential buffer
(b) Query plan optimization
(c) Unique identification of aggregate entries
(d) Cache eviction decisions

Correct Answer: (d)
Explanation: The metric entries are mainly used for cache eviction decisions. They are required to identify unused aggregate entries that only content the main memory.

3. Query Types
For which query types is the aggregate cache best suited?
Possible Answers:

(a) Recurring analytical queries
(b) Transactional queries
(c) All types of queries
(d) Distinct queries with aggregate functions

Correct Answer: (a)
Explanation: The aggregate cache speeds up the execution of analytical queries that are executed repeatedly.

Logging

1. Snapshot Statements
Which statement about snapshots is wrong?
Possible Answers:

(a) A snapshot is ideally taken after each insert statement
(b) A snapshot is an exact image of a consistent state of the database to a given time
(c) The recovery process is faster when using a snapshot because only log files after the snapshot need to be replayed
(d) The snapshot contains the current read-optimized store

Correct Answer: (a)
Explanation: A snapshot is a direct copy of the main store. Because of that, all the data of the database has to be in the main store when a snapshot is created in order to represent the complete dataset. This is only the case after the merge process, not after each insert, because inserts are written into the differential buffer and otherwise would be omitted.

2. Recovery Characteristics
Which of the following choices is a desirable characteristic of any recovery mechanism?
Possible Answers:

(a) Returning the results in the right sorting order
(b) Recovery of only the latest data
(c) Maximal utilization of system resources
(d) Fast recovery without any data loss

Correct Answer: (d)
Explanation: It is of course preferable to recover all of the data in as few as possible time. A high utilization of resources might be a side effect, but is not a must.

3. Situations for Dictionary-Encoded Logging
When is dictionary-encoded logging superior?
Possible Answers:

(a) If the number of distinct values is high
(b) If large values are inserted only one time
(c) If all values are different
(d) If large values are inserted multiple times

Correct Answer: (d)
Explanation: Dictionary Encoding replaces large values by a minimal number of bits to represent these values. As an additional structure, it requires some space, too. In return, huge savings are possible if values appear more than once, because the large values is only saved once and then referenced by the smaller key value whenever needed. If the number of distinct values is high or even maximal (when all values are different, which is also given if each value is inserted only once), the possible improvements can not be fully leveraged.

4. Small Log Size
Which logging method results in the smallest log size?
Possible Answers:

(a) Log sizes never differ
(b) Logical logging
(c) Common logging
(d) Dictionary-encoded logging

Correct Answer: (d)
Explanation: Common logging is wrong because it does not even exist. Logical logging writes all data uncompressed to disk. In contrast, dictionary-encoded logging saves the data using dictionaries, so the size of the data is implicitly smaller because of the compression of recurring values. Consequently, of the mentioned logging variants, dictionary-encoded logging has the smallest log size.

5. Dictionary-Encoded Log Size
Why has dictionary-encoded logging the smaller log size in comparison to logical logging?
Possible Answers:

(a) Actual log sizes are equal, the smaller size is only a conversion error when calculating the log sizes
(b) Because it stores only the differences of predicted values and real values
(c) Because of the reduction of recurring values
(d) Because of interpolation

Correct Answer: (c)
Explanation: Dictionary Encoding is a compression technique that encodes variable length values by smaller fixed-length encoded values using a mapping dictionary. As a consequences, it reduces the size of recurring values.

Recovery

1. Recovery
What is recovery?
Possible Answers:

(a) It is the process of recording all data during the run time of a system
(b) It is the process of cleaning up main memory, to "recover" space
(c) It is the process of restoring a server to the last consistent state before a crash
(d) It is the process of improving the physical layout of database tables to speed up queries

Correct Answer: (c)
Explanation: In case of a failure of the database server, the system has to be restarted and set to a consistent state. This is be done by loading the backup data stored on persistent storage back into the in-memory database. The overall process is called 'recovery'.

2. Server Failure
What happens in the situation of a server failure?
Possible Answers:

(a) The failure of a server has no impact whatsoever on the workload
(b) The system has to be rebooted and restored if possible, while another server takes over the workload
(c) The power supply is switched to backup power supply so the data within the main memory of the server is not lost
(d) All data is saved to persistent storage in the last moment before the server shuts down

Correct Answer: (b)

Explanation: A backup power supply is only a solution if there is a power outage but if a CPU or a mainboard causes an error, a backup power supply is useless. Saving data to persistent storage in the last moment before the server shuts down is not always possible, for example if there is a power outage and no backup power supply or if an electric component causes a short circuit and the server shuts down immediately to prevent further damage, there is no time to write the large amount of data to the slow disk. If a server has to shut down, incoming queries should be accepted and executed nonetheless. Therefore, the right answer is that the server has to be rebooted if possible and the data is restored as fast as possible. In the meantime the workload has to be distributed to other servers, if available.

Database Views

1. View Locations
Where should a logical view be built to get the best performance?
Possible Answers:

(a) close to the user in the analytical application
(b) in a third system
(c) in the GPU
(d) close to the data in the database

Correct Answer: (d)

Explanation: One chosen principle for SanssouciDB is that any data intensive operation should be executed in the database to speed it up. In consequence, views, which focus on a certain aspect of the data and often do some calculations to enrich the information with regard to the desired focus, should be placed close to the data. It this case, that means they should be located directly in the database.

2. Views and Software Quality
Which aspects concerning software quality are improved by the introduction of database views?
Possible Answers:

(a) Accessibility and availability
(b) Testability and security
(c) Reusability and maintainability
(d) Fault tolerance and usability

Correct Answer: (c)

Explanation: The introduction of database views allows a decoupling of the application code from the actual data schema. This improves the reusability and maintainability, because changes on the application code are possible without requiring changes to the data schema and vice versa. Additionally, existing application code can be used for many different data schemes, if the required schemas can be mapped via views to the data schema used by the application code. Availability, fault tolerance and testability are not improved by using views.

On-the-Fly Database Reorganization

1. Separation of Hot and Cold Data
How should the data separation into hot and cold take place?
Possible Answers:

(a) Application transparently, depending on workload relevance
(b) Automatically, depending on the state of the application object in its life cycle
(c) Block-wise, because data is already structured into horizontal blocks
(d) Manually, upon the end of the life cycle of an object

Correct Answer: (a)

Explanation: A separation of hot and cold data, harnessing application object and process information does not provide insights on the relevance of data. Closed invoices and cleared accounting documents might still be relevant for an application.

Mixed workload, enterprise applications demand a separation of hot and cold that is application transparent. This way, there is are no dependencies between application logic and data management procedures. Capturing data selection and projection statistics from the application workload provides the required flexibility to separate data into hot and cold.

2. Data Reorganization in Row Stores
The addition of a new attribute within a table that is stored in row-oriented format
...
Possible Answers:

(a) is very cheap, as only meta data has to be adapted
(b) is possible on-the-fly, without any restrictions of queries running concurrently that use the table
(c) is an expensive operation as the complete table has to be reconstructed to make place for the additional attribute in each row
(d) is not possible

Correct Answer: (c)

Explanation: In row-oriented tables all attributes of a tuple are stored consecutively. If an additional attribute is added, the storage for the entire table has to be reorganized, because each row has to be extended by the amount of space the newly added attribute requires. All following rows have to be moved to memory areas behind. Of course, the movement of tuples backwards can be parallelized if the size of the added attribute is known and constant, nonetheless is the piecewise relocation of the complete table relatively expensive.

3. **Cold Data**
   What is cold data?
   Possible Answers:

   (a) Data, which is still accessed frequently but is not updatable any longer
   (b) Data that is used in a majority of queries
   (c) The rest of the data within the database, which does not belong to the result of the current query
   (d) Data that is seldom or even never returned in query results

   Correct Answer: (d)
   Explanation: To reduce the amount of main memory needed to store the entire data set of an enterprise application, the data is separated into active (hot) and passive (cold) data. Active data is the data of business processes that are not yet completed and therefore stored in main memory for fast access. Passive data in contrast is data of business processes that are closed or completed and will not be changed any more and thus is moved to slower storage mediums like SSDs.

4. **Data Reorganization**
   The addition of an attribute in the column store ...
   Possible Answers:

   (a) slows down the response time of applications that only request the attributes they need from the database
   (b) has no impact on existing applications if they only request the attributes they need from the database
   (c) speeds up the response time of applications that always request all possible attributes from the database
   (d) has no impact on applications that always request all possible attributes from the table

   Correct Answer: (b)
   Explanation: In column-oriented tables each column is stored independently from the other columns in a separate block. Because a new attribute requires a new memory block, there is no impact on the existing columns and their layout in memory. Hence, there is also no impact on existing applications which access only the needed attributes that existed before.

5. Single-Tenancy
In a single-tenant system . . .
Possible Answers:

(a) power consumption per customer is best and therefore it should be favored
(b) all customers are placed on one single shared server and they also share one single database instance
(c) each tenant has its own database instance on a shared server
(d) each tenant has its own database instance on a physically separated server

Correct Answer: (d)
Explanation: If each customer has its own database on a shared server, this is the 'shared machine' implementation of a multi-tenant system. If all customers share the same database on the same server, the implementation is called 'shared database'. A 'single-tenant' system provides each user an own database on a physically separated server. The power consumption per customer is not optimal in that implementation of multi-tenancy, because it is not possible to share hardware resources. Shared hardware resources allow to run several customers on one machine to utilize the resources in an optimal way and then shut down systems that are not required momentarily.

6. Shared Machine
In the shared machine implementation of multi-tenancy . . .
Possible Answers:

(a) each tenant has an own exclusive machine, but these share their resources (CPU, RAM) and their data via a network
(b) all tenants share one server machine, but have own database processes
(c) all tenants share the same physical machine, but the CPU cores are exclusively assigned to the tenants
(d) each tenant has an own exclusive machine, but these share their resources (CPU, RAM) but not their data via a network

Correct Answer: (b)
Explanation: Please have a look at Sect. 32.3 on page 217.

7. Shared Database Instance
In the shared database instance implementation of multi-tenancy . . .
Possible Answers:

(a) each tenant has its own server, but the database instance is shared between the tenants via an InfiniBand network
(b) all tenants share one server machine and one main database process, tables are also shared
(c) the risk of failures is minimized because more technical staff (from different tenants) will have a look at the shared database

(d) all tenants share one server machine and one main database process, tables are tenant exclusive, access control is managed within the database

Correct Answer: (d)
Explanation: Please have a look at Sect. 32.3 on page 217.

Implications

1. Architecture of a Banking Solution
Current financial solutions contain base tables, change history, materialized aggregates, reporting cubes, indices, and materialized views. The target financial solutions contains ...
Possible Answers:

(a) only indexes, change history, and materialized views.
(b) only base tables, algorithms, and some indexes.
(c) only base tables, materialized aggregates, and materialized views.
(d) only base tables, reporting cubes, and the change history.

Correct Answer: (b)
Explanation: Because in-memory databases are considerably faster than their disk-focused counterparts, all views, aggregates and cubes can be computed on-the-fly. Also, the change history is dispensable when using the insert-only approach. Because main memory capacity is relatively expensive when compared to disk capacity, it is more important than ever to discard unneeded and redundant data. The base tables and the algorithms are still necessary, because they are the essential atomic parts of the database and indexes improve the performance while requiring only small amounts of memory.

2. Criterion for Dunning
What is the criterion to send out dunning letters?
Possible Answers:

(a) When the responsible accounting clerk has to achieve his rate of dunning letters
(b) Bad stock-market price of the own company
(c) A customer payment is overdue
(d) Bad information about the customer is received from consumer reporting agencies

Correct Answer: (c)
Explanation: A dunning letter is sent to remind a customer to pay his outstanding invoices. If a customer doesn't pay within the term of payment, he is overdue. Then a company has to send a dunning letter to call on the customer to pay, before it could demand an interest rate from him.

3. In-Memory Database for Financials

Why is it beneficial to use in-memory databases for financials systems?
Possible Answers:

(a) Because of the high reliability of data in main memory, less maintenance work is necessary and labor costs could be reduced.
(b) Easier algorithms are used within the applications, so shorter algorithm run time leads to more work for the end user. Business efficiency is improved.
(c) Operations like dunning can be performed in much shorter time.
(d) Financial systems are usually running on mainframes. No speed up is needed. All long-running operations are conducted as batch jobs.

Correct Answer: (c)

Explanation: Operations like dunning are very time-consuming tasks, because they involve read operations on large amounts of transactional data. Column oriented in-memory databases reduce the duration of the dunning run because of their extremely high read performance.

4. Languages for Stored Procedures

Languages for stored procedures are ...
Possible Answers:

(a) strongly declarative, they just describe how the result set should look like. All aggregations and join predicates are automatically retrieved from the database, which has the information "stored" for that.
(b) strongly imperative, the database is forced to exactly fulfill the orders expressed via the procedure.
(c) usually a mixture of declarative and imperative concepts.
(d) designed primarily to be human readable. They follow the spoken English grammar as close as possible.

Correct Answer: (c)

Explanation: Languages for stored procedures typically support declarative database queries expressed in SQL, imperative control sequences like loops and conditions and concepts like variables and parameters. Hence, languages for stored procedures usually support a mixture of declarative and imperative concepts and combine the best parts of the two programming paradigms to be very efficient.

Handling Business Objects

1. Object-Orientation

How does OO help in designing enterprise applications?
Possible Answers:

(a) OO concepts, such as encapsulation, aggregation, and inheritance help to design domain models, which can be discussed and validated with domain experts.

(b) OO allows domain experts to validate the data model of the application.

(c) OO provides the highest performance with regards to mathematical algorithms.

(d) OO provides a seamless integration of declarative concepts e.g., SQL.

Correct Answer: (a)

Explanation: Object-oriented concepts help system architects to design domain model, which reflect structures and relations of real world business entities. That way it is possible to validate semantic and logic of the algorithms and business objects with domain experts.

Bypass Solution

1. Transition to IMDBs

What does the transition to in-memory database technology mean for enterprise applications?

Possible Answers:

(a) The data organization will not change at all, but the source code of the applications has to be adapted

(b) Data organization and processing will change radically and enterprise applications need to be adapted

(c) There will be no impact on enterprise applications

(d) All enterprise applications are significantly sped up without incurring any adaptions

Correct Answer: (b)

Explanation: Traditional database systems separated the operational and the analytical part. With in-memory databases this is not necessary anymore, because they are fast enough to combine both parts. Furthermore, SanssouciDB stores the data in column-based format, while most traditional databases use a row-oriented format. Query speeds in enterprise applications will receive some improvements without any changes to the program code due to the fact that all data is in main memory and aggregations can be computed faster using column orientation. To fully leverage the potentials of the presented concepts, adaptions of the existing applications will be necessary so that the effects that the two mentioned major changes, the reunion of OLTP and OLAP and the column orientation, can be exploited in the program.

Glossary

ACID Property of a database management system to always ensure atomicity, consistency, isolation, and durability of its transactions

Aggregation Operation on data that creates a summarized result, for example, a sum, maximum, average, and so on. Aggregation operations are common in enterprise applications

Analytical Processing Method to enable or support business decisions by giving fast and intuitive access to large amounts of enterprise data

Atomicity Database concept that demands that all actions of a transaction are executed or none of them

Attribute A characteristic of an entity describing a certain detail of it

Availability Characteristic of a system to continuously operate according to its specification, measured by the ratio between the accumulated time of correct operation and the overall interval

Available-to-Promise (ATP) Determining whether sufficient quantities of a requested product will be available in current and planned inventory levels at a required date in order to allow decision making about accepting orders for this product

Benchmark A set of operations run on specified data in order to evaluate the performance of a system

Business Intelligence Methods and processes using enterprise data for analytical and planning purposes, or to create reports required by management

Business Logic Representation of the actual business tasks of the problem domain in a software system

Business Object Representation of a real-life entity in the data model, for example, a purchasing order

Cache A fast but rather small memory that serves as buffer for larger but slower memory

Cache Line Smallest unit of memory that can be transferred between main memory and the processor's cache. It is of a fixed size, which depends on the respective processor type

Cache Miss A failed request for data from a cache because it did not contain the requested data

Cloud Computing An IT provisioning model, which emphasizes the on-demand, elastic pay-per-use rendering of services or provisioning of resources over a network

Column Store Database storage engine that stores each column (attribute) of a table sequentially in a contiguous area of memory

Compression Encoding information in such a way that its representation consumes less space in memory

Compression Rate The ratio to what size the data on which compression is applied can be shrinked. A compression rate of 5 means that the compressed size is only 20% of the original size

Concurrency Control Techniques that allow the simultaneous and independent execution of transactions in a database system without creating states of unwanted incorrectness

Consistency Database concept that demands that only correct database states are visible to the user despite the execution of transactions

Cube Specialized OLAP data structure that allows multi-dimensional analysis of data

Customer Relationship Management (CRM) Business processes and respective technology used by a company to organize its interaction with its customers

Data Aging The changeover from active data to passive data

Data Layout The structure in which data is organized in the database; that is, the database's physical schema

Data Warehouse A database that maintains copies of data from operational databases for analytical processing purposes

Database Management System (DBMS) A set of administrative programs used to create, maintain and manage a database

Database Schema Formal description of the logical structure of a database

Deoxyribonucleic Acid (DNA) The carrier of genetic information

Dictionary In the context of this book, the compressed and sorted repository holding all distinct data values referenced by SanssouciDB's main store

Dictionary Encoding Light-weight compression technique that encodes variable length values by smaller fixed-length encoded values using a mapping dictionary

Differential Buffer A write-optimized buffer to increase write performance of the SanssouciDB column store. Sometimes also referred to as differential store or delta store

Dunning The process of scanning through open invoices and identifying overdue ones, in order to take appropriate steps according to the dunning level

Durability Database concept that demands that all changes made by a transaction become permanent after this transaction has been committed

Enterprise Application A software system that helps an organization to run its business. A key feature of an enterprise application is its ability to integrate and process up-to-the-minute data from different business areas providing a holistic, real-time view of the entire enterprise

Enterprise Resource Planning (ERP) Enterprise software to support the resource planning processes of an entire company

Entropy Average information containment of a sign system

Extract-Transform-Load (ETL) Process A process that extracts data required for analytical processing from various sources, then transforms it (into an appropriate format, removing duplicates, sorting, aggregating, etc.) such that it can be finally loaded into the target analytical system

Front Side Bus (FSB) Bus that connects the processor with main memory (and the rest of the computer)

Horizontal Partitioning The splitting of tables with many rows, into several partitions each having fewer rows

In-Memory Database A database system that always keeps its primary data completely in main memory

Index Data structure in a database used to optimize read operations

Insert-Only New and changed tuples are always appended; already existing changed and deleted tuples are then marked as invalid

Isolation Database concept demanding that any two concurrently executed transactions have the illusion that they are executed alone. The effect of such an isolated execution must not differ from executing the respective transactions one after the other

Join Database operation that is logically the cross product of two or more tables followed by a selection

Latency The time that a storage device needs between receiving the request for a piece of data and transmitting it

Locking A method to achieve isolation by regulating the access to a shared resource

Logging Process of persisting change information to non-volatile storage

Main Memory Physical memory that can be directly accessed by the central processing unit (CPU)

Main Store Read-optimized and compressed data tables of SanssouciDB that are completely stored in main memory and on which no direct inserts are allowed

MapReduce A programming model and software framework for developing applications that allows for parallel processing of vast amounts of data on a large number of servers

Materialized View Result set of a complex query, which is persisted in the database and updated automatically

Memory Hierarchy The hierarchy of data storage technologies characterized by increasing response time but decreasing cost

Merge Process Process in SanssouciDB that periodically moves data from the write-optimized differential store into the main store

Meta Data Data specifying the structure of tuples in database tables (and other objects) and relationships among them, in terms of physical storage

Mixed Workload Database workload consisting both of transactional and analytical queries

Multi-Tenancy The consolidation of several customers onto the operational system of the same server machine

Node Partial structure of a business object

Object-Relational Mapping (ORM) A technique that an object-oriented programm could use a relational database as if it is an object-oriented database

Online Analytical Processing (OLAP) see Analytical Processing

Online Transaction Processing (OLTP) see Transactional Processing

Padding Approach to modify memory structures so that they exhibit better memory access behavior but requiring the trade-off of having additional memory consumption

Prefetching A technique that asynchronously loads additional cache lines from main memory into the CPU cache to hide memory latency

Query Request sent to a DBMS in order to retrieve data, manipulate data, execute an operation, or change the database structure

Query Plan The set and order of individual database operations, derived by the query optimizer of the DBMS, to answer an SQL query

Radio-Frequency Identification (RFID) Wireless technology to support fast tracking and tracing of goods. The latter are equipped with tags containing a unique identifier that can be readout by reader devices

Real Time In the context of this book, defined as, within the timeliness constraints of the speed-of-thought concept

Real-Time Analytics Analytics that have all information at its disposal the moment they are called for (within the timeliness constraints of the speed-of-thought concept)

Recovery Process of re-attaining a correct database state and operation according to the database's specification after a failure has occurred

Relational Database A database that organizes its data in relations (tables) as sets of tuples (rows) having the same attributes (columns) according to the relational model

Row Store Database storage engine that stores all tuples sequentially; that is, each memory block may contain several tuples

SanssouciDB The in-memory database described in this book

Scalability Desired characteristic of a system to yield an efficient increase in service capacity by adding resources

Scale-Out Capable of handling increasing workloads by adding new machines and using these multiple machines to provide the given service

Scan Database operation evaluating a simple predicate on a column

Scheduling Process of ordering the execution of all queries (and query plan operators) of the current workload in order to maintain a given optimality criterion

Shared Database Instance Multi-tenancy implementation scheme in which each customer has its own tables, and sharing takes place on the level of the database instances

Shared Machine Multi-tenancy implementation scheme in which each customer has its own database process, and these processes are executed on the same machine; that is, several customers share the same server

Shared Memory All processors share direct access to a global main memory and a number of disks

Shared Table Multi-tenancy implementation scheme in which sharing takes place on the level of database tables; that is, data from different customers is stored in one and the same table

Single Instruction Multiple Data (SIMD) A multiprocessor instruction that applies the same instructions to many data streams

Software-as-a-Service (SaaS) Provisioning of applications as cloud services over the Inter- net

Solid-State Drive (SSD) Data storage device that uses microchips for non-volatile, high- speed storage of data and exposes itself via standard communication protocols

Speedup Measure for scalability defined as the ratio between the time consumed by a sequential system and the time consumed by a parallel system to carry out the same task

Stored Procedure Procedural programs that can be written in SQL or PL/SQL and that are stored and accessible within the DBMS

Streaming SIMD Extensions (SSE) An Intel SIMD instruction set extension for the x86 processor architecture

Structured Data Data that is described by a data model, for example, business data in a relational database

Structured Query Language (SQL) A standardized declarative language for defining, querying, and manipulating data

Table A set of tuples having the same attributes

Tenant (1) A set of tables or data belonging to one customer in a multi-tenant setup. (2) An organization with several users querying a set of tables belonging to this organization in a multi-tenant setup

Thread Smallest schedulable unit of execution of an operating system

Time Travel Query Query returning only those tuples of a table that were valid at the specified point in time

Transaction A set of actions on a database executed as a single unit according to the ACID concept

Transactional Processing Method to process every-day business operations as ACID transactions such that the database remains in a consistent state

Translation Lookaside Buffer (TLB) A cache that is part of a CPU's memory management unit and is employed for faster virtual address translation

Trigger A set of actions that are executed within a database when a certain event occurs; for example, a specific modification takes place

Tuple A real-world entity's representation as a set of attributes stored as element in a relation. In other words, a row in a table

Unstructured Data Data without data model or that a computer program cannot easily use (in the sense of understanding its content). Examples are word processing documents or electronic mail

Vertical Partitioning The splitting of the attribute set of a database table and distributing it across two (or more) tables

View Virtual table in a relational database whose content is defined by a stored query

Virtual Memory Logical address space offered by the operating process for a programm which is independet of the amount of actual main memory

Index

H. Plattner, *A Course in In-Memory Data Management*,
DOI 10.1007/978-3-642-55270-0, © Springer-Verlag Berlin Heidelberg 2014

Printed by Printforce, the Netherlands